T0201311

PROJECTS WITHOUT BOUNDARIES

Successfully Leading Teams and Managing Projects in a Virtual World

Russ J. Martinelli

James M. Waddell

Tim J. Rahschulte

Published by John Wiley & Sons, Inc., Hoboken, New Jersey

Published simultaneously in Canada

Library of Congress Cataloging-in-Publication Data is Available

ISBN 9781119142546 (Paperback)
ISBN 9781119376491 (ePDF)
ISBN 9781119376408 (ePub)

Cover Design: Wiley
Cover Image: © RamCreativ/iStockphoto

This book is printed on acid-free paper. ∞

Printed in the United States of America

10 9 8 7 6 5 4 3 2 1

CONTENTS

Life can be serendipitous at times. Most of the projects I have managed in my career as a new product development program manager have been virtual in nature and made up of geographically distributed teams. However, about five years after my last virtual project, I received an email requesting that I step in as the project manager for a project aimed at creating a secure cell phone for the government. The team, I was told, consisted of members of two organizations that had never worked together before, the software security group and the mobile devices group. Further, I was told that the team was highly distributed across the globe, with development centers in two locations in India, three locations in the United States, and one location in each of the countries of Ireland, Israel, and Germany.

The timing of the request is what was serendipitous. We had just begun the writing process for this book and were working through the primary differences between traditional and virtual projects. As I assumed my new virtual project manager role, the differences immediately began to emerge. So too did the various techniques for managing those differences. Personally managing a virtual, multinational, multicultural project while writing a book on the very same topic provided a wonderful opportunity to establish a practice-based foundation for the information found in the chapters that follow.

Part I stresses the importance of taking on the management of virtual projects with your eyes open. A keen awareness of the virtual project environment and the factors that create that

environment are more than *helpful* to a project manager, they are completely *necessary* to help create a new worldview. A virtual project environment is one that is characterized by separation of time and distance, by the inclusion of multiple national and potentially company cultures, and by a complete reliance on technology to facilitate team communication and collaboration. Part I brings forth the forces that drive the continuous increase in virtual organizations and projects, and the key differences between traditional and virtual projects that project managers must be aware of and use to their advantage.

All project managers must be prepared to assume two critical roles: being the manager of the project and being the leader of the project team. On a virtual project, there is often a shift in the balance of effort between these two roles. This shift in effort is caused by the distributed nature of the team, which demands significant focus on team leadership in addition to one's core project management responsibilities. Part II brings forth the key aspects of these two roles during the early stages of a virtual project. In particular, we describe crucial aspects of planning a virtual project (Chapter 2) while at the same time forming and building a high-performance virtual team (Chapter 3).

Part III continues the discussion on the two roles of the virtual project manager with focus on project execution (Chapter 4) and sustained virtual team leadership (Chapter 5). Attention then shifts to the importance of establishing a strong project network that connects the virtual team and enables the distribution of work, responsibility, and accountability on the part of those performing the work and empowerment to make necessary decisions locally (Chapter 6).

The final section of the book, Part IV, delves into a number of organizational factors that have to be established for a firm to experience sustained success in managing virtual projects. As a firm expands its project management activities to include international participants, the organization and its project teams become multicultural entities. Chapter 7 describes how national culture and company culture must converge to create the project culture and how virtual project managers must adjust their leadership tactics to account for multicultural factors.

Unlike co-located teams, virtual team members have to communicate and collaborate in a nearly exclusive asynchronous manner, and do so through the use of technology. Chapter 8 focuses on the primary role of technology for distributed project teams. We describe the various types of technology that are available to the virtual project manager today and then suggest a method for developing a strategy for choosing a suite of technological tools that will help the team overcome the challenges created by separation in time and distance.

Much of the pressure to succeed in managing virtual projects is unfairly placed on project managers. To address this situation, Chapter 9 presents the critical organizational factors that must be addressed to create a sustainable environment for virtual project success. These include instituting effective organization and team structures that foster collaboration and empowerment, changing recognition and reward systems to reinforce new behaviors and practices, and investing in new skills development for people thrust into the role of the virtual project manager.

Finally, a number of assessments are included throughout the book. Each assessment can serve as a survey, checklist, or tool to baseline and improve an organization's virtual project management and team leadership capabilities. The virtual project readiness assessment included in the Appendix will help an organization evaluate their readiness to enter the virtual project management arena, or to create a capability gap analysis and change transformation plan to increase their virtual project maturity. Virtual project readiness is assessed from organizational, team, and personal perspectives.

On behalf of the co-authors, our heartfelt thanks to the future readers of this book. I hope you find it both enjoyable and useful in your virtual project endeavors.

RUSS MARTINELLI

ACKNOWLEDGMENTS

We would like to thank the many people who have helped in making this book a reality.

To our experienced virtual project managers who graciously shared their expertise and experiences with us:

Janet Astwood

Sewa Bhatt

Stan Carr

Becky Christopher

Richard Cook

Ron Forward

Paola Genovese

Sylvie Huyskens

Hans-Juergen Junkersdorf

Sujith Kattathara Bhaskaran

Kris Knopf

Gideon Koch

Adit Liss

Kathy Milhauser

Amita Rao

Shlomit Shteyer

Victor Sohmen

Marc Valentin

To the team at John Wiley & Sons who continue to provide world-class support and guidance. In particular, we want to thank our executive editor, Margaret Cummins, our assistance editors Kalli Schultea and Amanda Shettleton, and our production editor, Seshadri Srinivasan. Your continued partnership and collaboration is greatly valued.

To our many colleagues and co-workers who have unknowingly contributed to the concepts presented in this work in many ways.

To our families who provide the support, encouragement, and patience necessary to complete the writing process, especially Andrea, Sue, and Doris!

We are truly blessed to be associated with such a wonderful and supportive community of people.

INTRODUCTION

WORKING IN A VIRTUAL WORLD

As Jeremy Bouchard adjourned his weekly team meeting, he paused to reflect how much his project environment had changed in less than six months. Until that time, the projects that he managed were traditional in that the project team was co-located, allowing the team members to conduct their team meetings across the table from one another and Jeremy to "manage by walking around" on a daily basis. Today, as he adjourned the meeting while sitting alone in his office staring at his computer screen, he realized how drastically things had changed now that he is the manager of a virtual project.

Jeremy's story is one of sudden change—change that was driven by the acquisition of his company by a much larger company with a global presence, Sensor Dynamics, a manufacturer of specialized sensor products in an emerging technology segment called the Internet of Things. Unlike many of his colleagues, he welcomed the change and looked forward to applying his well-honed project management skills on a larger scale with Sensor Dynamics.

That opportunity came quickly. Jeremy was assigned the project manager role for a new human biometric sensor product—an emerging market with rapid growth potential. Through a recent company reorganization, which is common following an acquisition, Jeremy is now reporting to the Project Management Office director, a veteran employee of Sensor Dynamics. His project team is a combination of people from his old organization and the new one. They are distributed across three locations in his home country and three locations in other countries.

He now finds himself leading a team of people, most of whom he has never met personally. Six weeks into the project planning process, Jeremy is trying to come to terms with the increased difficulty associated with managing a virtual project versus a traditional project. As he says, he is feeling like a "fish out of water" while trying to learn the nuances associated with changes in common project management practices and the complexities associated with leading a distributed, and mostly virtual, project team.

Even the most common project management tasks, like creating the project charter, are proving to be monumental challenges. In particular, Jeremy has continual disagreements regarding team member roles and responsibilities. Despite repeated attempts, he has not been able to establish team consensus. Additionally, there is growing conflict between two key project team members on the goals of the project. The conflict is threatening to cause wider team dysfunction. Because the individuals are separated by geographical distance, the conflict is escalating in every email exchange between the two.

Jeremy is also learning about people's reluctance to collaborate with one another on a distributed team. He has tasked two team members to develop a combined task plan since their deliverables will be intertwined. Two weeks into the effort, it has become apparent that they have not yet begun to communicate, let alone collaborate in any way to create the task plan. Team members seem very reluctant to share information. Jeremy cannot determine if the problem is a lack of trust or if there is an underlying sense that "information is power" to the owner. Hence, they are keeping information to themselves.

Then there is the technology problem. Jeremy has had to revert to the use of phone conversations and email in order to communicate and collaborate reliably. Even though Sensor Dynamics has deployed an enterprise-level team collaboration system, some team members are either unable or unwilling to adapt to the technology. This is especially true of team members in countries other than Jeremy's.

The most frustrating thing to Jeremy, however, is the realization that management by walking around is now impossible. He has not been able to establish a new method for connecting with his team members or for staying on top of project progress.

Jeremy decided to raise his issues with his manager, Brent Norville. Norville, the Project Management Office director of Sensor Dynamics, has been with the company for over a decade and has experienced the transformation of the company to a virtual organization firsthand. As Bouchard and Norville began their conversation, Bouchard shared that he was having trouble adjusting to the virtual project environment he was now working within.

Norville explained that he understood that the virtual project environment in which Sensor Dynamics executes its projects is significantly different than what Bouchard was used to. He also explained that he understood that the sudden change from traditional to virtual project management is the exception rather than the norm. Sensor Dynamics as a company has been transitioning for more than a decade, and most project managers who come into the organization have had some experience working on or managing virtual projects. Norville explained that it takes time to understand and effectively work in a virtual project environment. He also explained that many of the factors that make managing a virtual project so different have little to do with the project management fundamentals that Bouchard is well versed in. Rather, the differences come in understanding how those fundamentals have to be practiced differently and

how more focus, time, and personal effort have to be applied toward leading the virtually distributed team. Norville also offered to act as a coach to Bouchard when needed to accelerate his transition from traditional to virtual project manager.

This conversation led Bouchard to realize that he was playing a game of catch-up to many of his project management peers who had at least some experience managing a virtual project and leading a distributed team.

Now, we have to recognize that Bouchard's story is an extreme example. Fortunately, the majority of project managers are not introduced to the world of virtual project management in such a sudden and abrupt manner. That does not mean that we did not each experience all or many of the perplexing problems facing Bouchard. We more than likely encountered them over time instead of all at once. Much like wading from the shallow end to the deep end of a swimming pool when learning to swim, most project managers can transition from traditional to virtual project management practices at a measured pace as their virtual awareness and confidence increases. However, we still hear stories of people being thrown in the deep end of the pool and struggling to learn and apply best practices to be effective.

Truth be told, nearly all projects today are at least partially virtual in nature. If your company outsources some of its work, or allows employees to telecommute, or is distributed in multiple locations (even in the same city), you are working for a virtual organization. Of course, distributed team members and the work they perform is not new, but to view our companies as virtual organizations is a paradigm shift for many. Even teams that are co-located work somewhat in a virtual manner through the prolific use of email, instant messaging, collaboration sites, social media technologies, and other forms of mobile applications. How often have you sent an email or instant message to people on your team whose offices are in the same building or possibly right next to your own? Likewise, how many times do we engage in teleconferences where we can hear a person speaking who sits near us in one ear and then a few milliseconds later in our other ear through the telephone receiver? For some, like Nora Bennington, this is a strange new world:

> I just don't understand it sometimes. I'm constantly getting IMs [instant messages] from people sitting no more than 30 feet from me, wanting to engage in a conversation on a particular topic. When I get up and walk over to their offices to have a real conversation, they react with complete surprise. Like I'm violating some unwritten policy that we can't engage in real conversation anymore.

For Bennington and others, getting used to working on a virtual project is a slow process. Some don't even realize that the project world has changed so rapidly around them. In September 2015, Global Workplace Analytics, a company that helps organizations understand emerging workplace strategies such as telecommuting, open office, and flexibility work, updated its statistics on what it calls distributed or mobile work in the United States. It is showing some significant growth in this measure. From 2005 to 2014, this demographic of the workforce doubled from 1.8 million to 3.7 million. This statistic includes both nonprofit and profit-based organizations.[1]

But what defines a virtual project? By itself, the use of technology to communicate and even collaborate does not define a virtual project. Rather, a virtual project is one in which its resources are separated by geographic or temporal space.[2] In extreme cases, the members of a virtual project are separated by organizational boundaries, national borders, continents, and multiple time zones.[3] In such situations, it is highly likely that the members of a project team will never meet face-to-face. For many of us, this has been a major shift in the way we participate on project teams. For others, especially those who entered the workforce over the past 10 to 15 years, project work has always been conducted virtually. Within the next decade, the topic of virtual projects and virtual teams likely will no longer garner such attention, just as topics

such as project scope and the triple constraints have moved from interesting to sleeper topics. Managing virtual projects will be ingrained in the way we do business. Until then, however, many project managers will still experience a transition from the practices of managing traditional projects to new and modified practices required to manage virtual projects. The transformation will cause us to redefine our companies and the projects within our companies as collaborative systems with networked structures, and work outcomes that are not built on organizational hierarchy but on trust, relationships, and communication.[4] Integration and collaboration are now more than technological capabilities; they are central to how virtual project work is performed.

The purpose of this introductory chapter is to broaden awareness of the factors that contribute to the creation of virtual organizations and subsequent virtual projects, expose the primary differences between traditional and virtual projects, and help accelerate people's transition from being effective traditional project managers to virtual project managers.

Forces Driving Virtual Transformation

A common question being posed by many project managers is: "Why does the pace of transformation to this new virtual project paradigm seem to be accelerating?" The reasons are important for project managers to understand because the transformation to virtual projects is testing the viability of many traditional project management practices and methodologies. Further, the answer to that question does not lie within the world of project management. Rather, the accelerated pace of the transition to virtual projects is being driven by the globalization of our economies and our businesses.

As companies participate in the global marketplace, business operations (including project operations) expand beyond their corporate boundaries. In 2009, the Economist Intelligence Unit, an organization that provides executives with practical business information on macroeconomics, conducted a survey of executives to evaluate the extent to which companies in Europe are using virtual teams. The survey included 407 firms from various industries with annual revenues of greater than $100 million. Of the survey participants, 78% indicated that they use virtual teams in their firms. The survey authors concluded that working in virtual teams is growing and that the majority of the business executives surveyed are positive regarding their use of virtual teams to perform the work of their firms. The authors also indicated that the use of virtual teams has enabled these firms to gain access to a global talent pool and has been a factor in improving their organizations' performance against their competitors.[5]

This expansion requires everyone within an organization to develop a broader view of the environment in which businesses operate. This is particularly true of project managers, who are on the front lines of globalization. Project managers must develop a worldview—an awareness of the business environment outside of their own region, industry, and country that includes social, economic, and political factors and trends that can affect the businesses they work for. It is from a worldview that managers can develop an understanding of how economic, political, and technological forces that are driving today's global marketplace interact, how that interaction creates new strategic opportunities, and how those new strategic opportunities lead to the virtualization of projects. We call these forces the *globalization forces*.

Knowledge of the three primary forces—economic, political, and technology—provides virtual project managers a greater context of the dynamics in play within their project environments. This greater context and awareness is important because it frees project managers from feeling as if the virtual challenges they may be experiencing are a result of poor decisions on the part of their senior corporate leaders or of their own inabilities to manage a virtual project effectively. Instead, the broader awareness helps managers realize they are now part of a very dynamic business environment that is being played out on a global scale.

Economic Forces

The basis of global economics involves the creation of economic interrelations across geographical boundaries as defined by national borders through the production, exchange, and consumption of goods and services. Global free-market economics is stimulated by the flow of money and capital between nations by large and small transnational corporations, international economic institutions, and trading systems that create interdependencies between national economies.[6]

World economics of the past several centuries has been dominated by two philosophies: free-market economics and Keynesian economics. Free-market economics is rooted in the view of Adam Smith (1723–1790), who defined markets as self-regulating mechanisms that drive toward a balance between supply and demand of goods and services.[7] Within a free-market system, trade in goods and services between nations is unhindered by government-imposed restrictions such as taxes, tariffs, and quotas. Free-market economics is characterized by free access to markets, free movement of labor between nations, and free movement of capital between nations.

Keynesian economics, conversely, advocates nation-state influence of world economic policy. John Maynard Keynes (1883–1946) believed that economic systems would not automatically balance by themselves; therefore, macroeconomic control by government institutions is needed to ensure balance and equity within an economy. Macroeconomic control includes control of money supply, control of interest rates, and control of market access. Keynesian theory recognizes that economic systems will realize points of downturn and even depression and that these systems are not self-correcting; rather, they need support and influence from government interventions to boost the system in recovery.[8]

Whether dominated by free-market policy, Keynesian policy, or a combination of the two—today's most common method—economics is the primary driver that motivates corporate leaders to explore beyond their traditional strategic boundaries. It is economics that drives the world's entrepreneurs and business leaders to seek new markets for their goods and services, to find new suppliers for their raw materials, to develop worldwide sources for production and distribution, and generally to evaluate the world's resources for potential competitive advantage and product optimization. Economics therefore is the driving force that creates our virtual organizations and virtual project work.[9]

Political Forces

World politics is the second primary force that affects globalization. Rarely have economic globalization forces been able to operate independently from political forces. Most often, global economic expansion and contraction is set in motion by a series of political actions. The basis of world politics is the generation, distribution, and control of power and influence.[10] For many centuries, control of power has been achieved by creating territorial lines that defined national borders. In doing so, artificial boundaries have been created that allow us to view the world as a series of "domestic" and "foreign" relationships.

The political force pressuring globalization involves the partial permeation of these national boundaries in order to expand the trade of goods and services. Fledgling entrepreneurs have not been able to achieve expansion of their businesses on a global basis without the support of their governments and of the governments of their trading partners.[11]

Although recently we have seen the world influenced by the decisions of the Global20 nations, over the past 50 years, governments have funded the early development of technologies that were later commercialized and are now common in our personal and work lives today. Many of these advancements came out of the competition and conflict between the U.S. and Russian governments in trying to win the race to the moon and the Cold War. Today we are witnessing competitive business wars beyond Russia and the United States. Businesses from around the world are competing to be first to market with a sustainable product base and growing customer demand. Those with the most compelling offerings and most effective globalization strategy/execution combination will win, and the followers will be forced to resort to reactive strategies for survival. Because of these competitive conditions, political forces directly affect globalization and virtual project work.[12]

Technology Forces

Technology is the third primary globalization force. Although economics is the true driving force for globalization and politics is mainly a guiding force that either stimulates or contracts globalization, technology is the force that makes globalization both more effective and efficient. Said another way, the *speed* of globalization is dependent on the conditions for technological use and advancement of technology development.

The basis of technology as a globalization force is in the development and dissemination of new ways to expand our global reach, to facilitate the interaction and interdependencies of humans across the globe, and to enable the flow of monetary exchange across national borders.[13]

Early technology development focused on more effective forms of transportation to help explorers overcome geographical barriers that prevented them from opening new trade routes to expand their markets. Later, new power technologies, such as coal, steam, and petroleum, helped to make transportation of goods and services much more efficient.[14] This led to the invention of mechanized shipping, railway systems, and automotive and air transportation. Additionally, the introduction of electricity spawned new communication technologies, such as the telegraph, radio, television, telephone, and electronic money exchange.

Today, much technological development has been focused on the introduction of collaborative technologies that have resulted in such deep permeation of national boundaries that those boundaries no longer prevent people from collaborating and participating in the exchange of goods and services. These technologies include internet technologies, business-to-business technologies, and work-flow technologies that enable knowledge work to be disaggregated, distributed, and reintegrated across the globe. Collectively recognized as "technology," this force speeds the rate at which globalization can expand, and it also accelerates the potential of virtual project teams.[15]

Interaction among the Globalization Forces

It helps to look at each of the three primary forces of globalization separately to better understand their influence on globalization. However, the forces themselves do not operate independently. It is the interaction among economic, political, and technology forces that has historically had the most dramatic influence on globalization.

We use the tricircle model shown in Figure 1.1 to graphically demonstrate the interactions among the globalization forces and the resulting impacts on the world economies. We provide this analysis to help virtual project managers become more aware of the dynamic forces in play within the environment in which their projects operate.[16]

Globalization can be characterized by drivers, enablers, and accelerators. Economics is the globalization *driver*, meaning the quest for greater economic gain has fueled the human desire to connect with others across the globe to expand the production and sales of goods and services primarily for prosperity, but also for human connection.

Politics is the globalization *enabler*. Political policy is driven by the agendas of the world's nation-state leaders, which in turn either positively or negatively affect global economic interconnection between nations.

The third force, technology, is the globalization *accelerator*. Historically, significant advances in various technologies have increased the pace in which people and economies have become interconnected.

These three forces are not static. Rather, each is very dynamic and always in flux. When the globalization forces are independent in nature, as demonstrated in Figure 1.1, it represents a period of slow globalization expansion or, more likely, globalization contraction. When the globalization forces become highly integrated, as demonstrated in Figure 1.2, a state of globalization exists where all three forces are at work to facilitate the wide and rapid expansion of globalization. Such is the state of globalization today, where most world economies

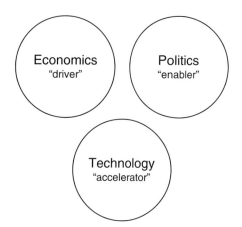

Figure 1.1: Primary Globalization Forces

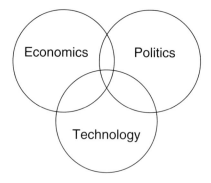

Figure 1.2: Integration of the Globalization Forces

and monetary exchanges are driving globalization, political stability, and alignment are enabling continued globalization into new and larger markets, and the advent of new work-flow technologies has accelerated knowledge work activities, allowing work to be distributed digitally to workers across the globe.[17]

The integration of the globalization forces is causing changes that will continue for the foreseeable future despite continual resistance to the trend as well as constant challenges associated with executing in a global environment. For those of us caught in the changing tide, it is time to adjust our perspectives and sharpen our skills to ensure our personal success—and the success of our companies—in this wave of globalization that is fueling the virtualization of our companies and our projects.

Rise of Virtual Organizations and Projects

For a large number of companies, participating in the global marketplace has become a matter of survival and sustainability. In order to compete in a global marketplace, these companies have had to develop new business strategies that break traditional organizational and strategic boundaries. In the process, virtual organizations and virtual projects are formed, and new capabilities and tools are developed that enable virtual project work to be performed.

The central purpose of most enterprises is to provide value to their stakeholders. Whether an enterprise has a mission to make a profit for its shareholders or to reinvest it profits in services that benefit its customers or clients, many senior enterprise leaders find it necessary to compete in a global marketplace in order to continuously create and deliver value. To remain relevant in this game of value creation, senior leaders have to look beyond their organizational, company, geographical, and cultural boundaries when establishing future business strategy. Today, corporate leaders need to think in terms of *strategic* boundaries, not *physical* boundaries.[18] If their competitors are playing on a global scale, so must they.

Crossing strategic boundaries means taking actions such as acquiring other organizations (or allowing themselves to be acquired), developing strategic alliances with partners that complement and expand their current capabilities, outsourcing some of their processes to outside firms that can perform the work more efficiently, moving portions of their operations to new locations to reach new markets, and looking abroad to acquire talent outside one's home location. Any of these strategic actions can immediately create a distributed organization or further expand an already distributed one. In the process, a virtual organization is created or expanded. For an example of how strategic business actions can create a virtual organization, see the box titled "Virtual Telecom."

Virtual Telecom

With security breaches continuing to climb, Juniper Networks realized it lacked key security protection capabilities in its products that threatened the future of its product lines. Company leaders developed a strategic goal to bolster the security capabilities of company products within the next two years and spawned a discussion of whether to develop the needed capabilities or buy them.

In 2005, the California-based Juniper made the strategic business decision to acquire a company in Massachusetts named Funk Software in an effort to quickly solve the security problem and integrate the newly acquired capability into its products. Up to this point, security capabilities were primarily developed in-house. The critical time goal was the variable that caused Juniper senior leaders to cross traditional strategic boundaries and acquire Funk Software.

As a result of this strategic decision, Juniper became a virtual enterprise consisting of organizational entities on the West Coast and East Coast of the United States. The decision to integrate the newly acquired network security capability into its products had an immediate impact on a number of Juniper project teams as well. All projects involving the security capability became virtual in nature, with resources and team members suddenly separated by 3,000 miles and three time zones.[19]

As this example shows, strategic business decisions can expand a company beyond its traditional organizational boundaries and, in the process of doing so, create a virtual organization. In like manner, since project structure and composition directly mirrors organizational structure, these same strategic business decisions create virtual projects that also lack traditional boundaries.

With a political and business environment that supports the expansion of enterprises to nearly all geographies of the world, physical location is no longer a constraining factor to creating and implementing business strategies. Because of this, the virtual organization is rapidly evolving to be the new norm.

As companies redefine themselves by optimizing the implementation of their strategies across company and geographical boundaries, it has a direct effect on their projects. As described in the "Virtual Telecom" example, the virtualization of projects can be immediate and sudden. This is why many project managers are surprised by the new virtual paradigm shift and find themselves inadequately prepared even though they have honed their project management knowledge and have years of experience. Because virtual projects have some significant differences associated with them, management of virtual projects requires retooling our project management practices, processes, tools, and skills. In some cases, the differences just require project managers to refocus on practices, processes, and tools for which they have been trained but that take on a higher degree of importance for virtual projects. Two examples are project chartering and clearly documenting project team members' roles and responsibilities. In other cases, the differences may require new practices,

processes, and tools, such as influencing virtual stakeholders and using collaboration technologies. Before project managers can make adjustments to their practices, processes, and tools, they need to understand the primary factors that make managing virtual projects so different.

Virtual Projects Are Different

There are, in fact, *many* differences between virtual projects and traditional projects. Attempting to discuss all the differences would be overwhelming. However, a number of significant differences create major changes in the role of project managers and how they manage a virtual project differently from a traditional project. The differences are evident in both the management of the project management processes and in the leadership of the project team. The key differences that create the most impact to the management of virtual projects include:

- Distribution of project team members
- Higher level of complexity to contend with
- Greater focus on integration of work
- Distributed decision making
- Greater hesitance to share information
- Difficulty in maintaining alignment to strategic goals
- Difficulty establishing cross-team connections
- Greater reliance on technology for communication and collaboration
- Greater challenge to monitor and control project work

Virtual Project Teams Are Distributed

The most obvious difference between virtual projects and traditional projects is the fact that team members are geographically distributed on virtual projects. Our first attempts at leading virtual projects normally involve trying to replicate the team and resource management practices used for traditional teams in the virtual team environment with little consideration for the effectiveness of the fit.[20] This approach usually faces challenges because different approaches are required in the virtual project environment for team building, communication, collaboration, and integration of distributed work. This is not to say that all traditional practices and processes for managing a project have to be modified for virtual projects. As we explain in chapters to follow, the key is knowing which can carry over, which need to be modified, and which need to be replaced with new practices.

The overwhelming amount of literature and training over the past 15 years on the subject of virtual projects has focused on the people side of project management. This is probably due to the fact that historical approaches to training and certifying project managers has left a vast gap in knowledge required for managing the people side of projects. As project managers transition from managing traditional projects to managing virtual projects, people issues become amplified because they affect team cohesiveness and trust between team members. The people issues then in turn affect how well traditional project management methods and processes work on a virtual project.

In the chapters that follow, we focus on the necessary practices for building a virtual team with a sense of common community and purpose, making changes necessary for increased team monitoring and feedback, managing across multiple time zones, dealing with virtual conflict and differences in language and culture, communicating asynchronously via technology, and changing reward systems that are necessary for geographically distributed project teams.

Virtual Projects Are More Complex

Virtual projects are built on interconnectivity of organizational, human, and electronic networks. This high level of interconnectivity makes virtual projects more complex by nature than traditional projects.

After years of working with complexity, Richard Cook, the deputy project manager of the Mars Science Laboratory at NASA, concludes that the word *complexity* "is frequently thrown around as a sort of synonym for difficult."[21] Cook notes, however, that "complexity is the quality of being intricately combined." He distinguishes complexity from *difficult* based on "the number of interconnected elements that are tied together technically, programmatically, and organizationally."

This gives us great perspective on why virtual projects are so complex. Virtual project teams perform much of their communication and collaboration through a highly interconnected technology platform. Further, the various outcomes and project deliverables generated by the distributed team are highly interdependent and need to be integrated programmatically to create a holistic solution. Finally, the ability to distribute project work across the globe opens the opportunity for interconnected collaboration between partner organizations. The more distributed the virtual project, the more complex it becomes, as illustrated in Figure 1.3.[22]

When project teams are co-located in a single location—Palo Alto, California, in our example—the workplace is physical. Even though some elements of the project have virtual characteristics (such as electronic communication), complexity is strictly related to the project structure itself.

If we look at the next logical step in creating a virtual organization, expanding nationally, complexity associated with working in a virtual environment becomes additive to the base complexity of the project. Now interconnectedness becomes separated by time and distance and must be held together by human, organizational, and technological networks.

Since international boundaries are no longer a constraint to business expansion and partner

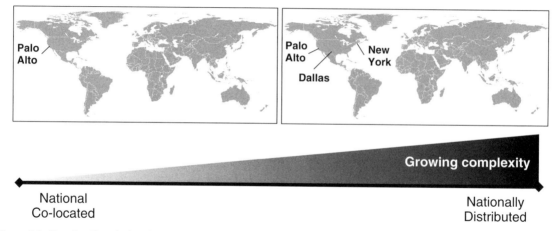

Figure 1.3: Growing Complexity with Added Distribution

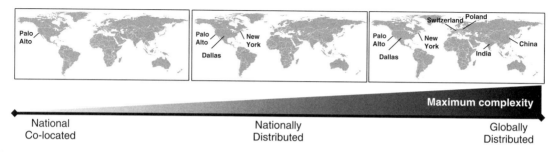

Figure 1.4: Maximum Complexity with Global Distribution

alliances, when organizations expand their virtual organization internationally, project complexity once again increases, as depicted in Figure 1.4.

In globally distributed virtual projects, additional cultural, language, and time zone challenges emerge. These factors create an environment of maximum project complexity that has to be comprehended and managed.

The criticality of performing a project complexity assessment is a distinguishing difference between virtual and traditional projects. A complexity assessment (Chapter 2) should be performed early in the life cycle of a virtual project. The information gained from the assessment will assist project managers in determining the level of complexity of their projects. The information also aids project managers in determining the skills and experience levels required of project team members and in guiding the implementation of key project processes, such as change management and risk management, evaluating the amount of contingency reserve to incorporate into the project schedule and budget, and adapting their management style to the complexity level of the project.[23]

Greater Focus on Integration Required

As complexity increases, the need for more work interdependencies between virtual project team members also increases. Managing a virtual project means designing and managing a network of interdependencies among distributed team members.

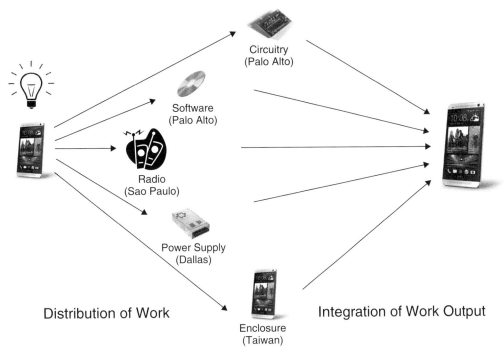

Circuitry
(Palo Alto)

Software
(Palo Alto)

Radio
(Sao Paulo)

Power Supply
(Dallas)

Enclosure
(Taiwan)

Distribution of Work   Integration of Work Output

Figure 1.5: Distribution and Integration of Virtual Work

As project work becomes more distributed, it becomes more decentralized. This is illustrated in Figure 1.5, which shows a simplistic view of the distribution of work packages for a globally distributed project team chartered with designing and developing a mobile device, such as a smartphone.

Not only does the product have to be designed from a systems perspective, so does the distribution of work to create the product. The illustration indicates that the project work is accomplished in a very fragmented manner on a virtual project, and it will remain fragmented unless the integration of the work outcomes of the various project specialists is purposefully managed and integrated.

Of the nine knowledge areas associated with the Project Management Institute's project management methodology, the integration knowledge area often receives the least attention and focus from project managers. This is understandable for traditional projects because much of the integration

of project work is accomplished through the centralized management of tasks and interdependencies. When tasks become highly distributed, which is the case on virtual projects, it becomes challenging to maintain centralized management at the task level. Most of this responsibility is in fact decentralized and distributed to the various locations where the work is being performed. Project managers now must focus on establishing centralized management of the project outcomes or deliverables. An important element of centralized management at this higher level is the integration of the various project outcomes to create a synthesized and holistic solution.

This integration of work is a difficult process for many reasons, one of which has to do with testing base assumptions made by project team members as they conduct their work. Like members of traditional projects, virtual project team members have to perform much of their work, and create their project deliverables, on a series of assumptions. When incorrect assumptions are made, they are normally discovered and corrected through direct communication and collaboration with other team members. On virtual projects, however, there is a decrease in direct communication among team members, a lack of informal meetings, and limited synchronous collaboration as compared to co-located teams. It is rare, therefore, that incorrect assumptions are discovered through normal means of communication and collaboration. Rather, they usually are found when work outcomes and deliverables are created and then integrated with other elements of the project. Incorrect assumptions emerge in the form of inconsistencies that are discovered between what was created and what was expected by other team members. These inconsistencies then have to be reconciled through the integration process.

Decision Making Is Decentralized

There are hundreds, some say thousands, of decisions that have to be made during the course of a project. Some are small and relatively insignificant, some are big and critical to the success of the project, some are incremental and follow a pattern or trend, while some are ambiguous and have never been encountered before. Whether big or small, nearly all decisions need to involve discussion, debate, and input from various project players. Traditional projects benefit from co-located team members and stakeholders who can assemble and engage in a rich discussion concerning a particular decision. Virtual teams do not have this luxury, and the separation of team members by time and distance can become an inhibitor to timely decision making. Therefore, a common ailment affecting virtual teams is slow decision making.[24] To combat this ailment, modifications to project management decision processes have to be made.

The team charter (Chapter 2) is an ideal project artifact for documenting team decision-making processes.[25] In particular, project decision processes are likely to be different in the virtual project environment as compared with the more traditional project environment that most team members have experienced.

On a traditional project, the project manager is the primary person providing leadership for the project. On a virtual project, however, leadership typically is shared among team members based on location and task at hand. This includes decision-making leadership.[26] A more complex centralized/decentralized decision framework has to be established for virtual projects. (See Chapter 6.) Decisions that directly affect the success of the project, such as those that can change the project schedule, need to remain centralized with the project manager. Other decisions need to be decentralized and moved to where the decision outcome will be implemented. These decisions, such as the hiring of a particular project team member, become the responsibility of the project personnel within the location who can make the decision in a more timely manner.

In order to make distributed decision making possible, two key factors must be present. First, the project manager must assign and document specific decision responsibility to the distributed project team members. Second, distributed decision making can quickly become ineffective if those given decision responsibility are not also given the authority to make and own the decisions. Project managers must *empower* the virtual team members who have been delegated decision responsibility by clearly communicating and documenting their decision authority to all project stakeholders.

Greater Hesitance to Share Information

Because virtual project team members share most of their information electronically, often they hesitate to share information at all. This is especially true early in team formation and engagement. The problem has much to do with trust, particularly trust that the information will be used properly. Unfortunately, project decisions, like all decisions, rely on information. Decision makers on virtual projects have to be more deliberate in requesting, extracting, and brokering the exchange of information that supports their decisions than their counterparts on traditional projects, where the flow of information is more forthcoming, fluid, and facilitated by verbal conversation.

Virtual project managers also have to be more systematic about the collection of information and repeatedly ask team members if they have anything else to share that hasn't already been supplied. Hoarding information as a source of power is a common phenomenon, and it can be a barrier to success for a virtual project. It is vital for virtual project success that project managers be proactive in searching for information and not passively waiting for it to arrive; it may never be supplied.

Maintaining Strategic Alignment Is Difficult

For any project, success is ultimately measured in some form of return on investment (ROI).There are many ways to measure ROI, but they all boil down to a common formula:[27]

ROI = Identification of strategic goals versus ability to execute

It is important to note that maintaining alignment to the strategic goals for which the project was intended to achieve is in many ways a much more difficult task for virtual projects. Team members on a virtual project are more geographically, physically, and sometimes culturally isolated not only from other team members but from their organization as a whole.[28]

It is the responsibility of the managers of virtual projects to ensure that all team members share a clear sense of how their work fits in with the overall project vision and that they are committed to the strategic success of the organization. Maintaining strategic alignment begins and ends with clear communication on the part of project managers on two primary pieces of information:

1. Where we are going (strategic goals to achieve)

2. How we are going to get there (collaborative planning and execution)

This information is essential for ensuring that the distributed work outcomes and actions taken do not violate a strategic principle or interfere with the strategic direction of the organization. If a violation occurs, there is an increased likelihood that execution outcomes will become misaligned with the strategic goals of the business.

Like traditional projects, virtual projects work within the functional paradigm where team members report to a functional department, or silo, and are "loaned" to the project on a temporary basis. This arrangement can interfere with team members' commitment to project goals as they now have two alliances: one to the functional organization to which they belong and one to the project. Functional work obligations often assume a higher priority, causing team members to lose sight of the project's strategic objectives and how their piece of the project contributes to the overall desired outcome.[29] This is a cultural issue more than a project issue, but it is left to project managers to help team members navigate these dual priorities by establishing a clear and common purpose. The project charter, team charter, project vision, and success measures take on a more critical role on virtual projects and become the most valuable tools for virtual project managers in establishing and maintaining alignment to the strategic goals driving the need for the project.

Cross-Team Connections Are Slow to Develop

We have already explained that virtual projects are primarily established on a series of networks (organizational, technological, and human) and not physical locations. This additional complexity, of course, creates challenges because very few network connections between team members exist on a virtual project, especially in the early stages of team formation. The human network especially takes time to establish because connections are built on trust, personal relationships, and direct communication between team members.

Until network connections between virtual team members are fully in place, it falls on project managers

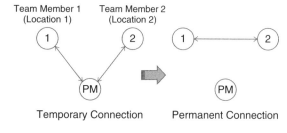

Team Member 1 (Location 1)   Team Member 2 (Location 2)

Temporary Connection   Permanent Connection

Figure 1.6: Communication and Collaboration Conduit

to personally establish many of the cross-team connections. Project managers find themselves having to directly and purposefully facilitate the communication and collaboration among team members until a point where the networks are sustainable. (See Figure 1.6, where *PM* stands for project manager.)

The need to facilitate connections is the case on traditional projects as well. The difference is that once a network connection is established by project managers of traditional projects, the ability to hold face-to-face interaction quickly takes over and solidifies the connection. On virtual projects, team member interaction is occurring through the use of technology without the benefit of face-to-face interaction. As a result, project managers have to stay involved in the interaction through a larger number of communication exchanges and, therefore, for a longer period of time. Sewa Bhatt, an experienced virtual project manager, explains her personal experience with establishing team connections:

> One thing that has helped me with my virtual teams is making special effort to establish the connections—which also includes social connections and networking. This seems to happen naturally with co-located team members, but not with virtual teams.

Having to act as conduits for cross-team communication and collaboration puts additional pressure on project managers—pressure that demands significantly greater commitment of their time. Project managers new to managing virtual projects often are unprepared for the amount of time and effort required to establish the cross-team network. However, it is a critical role that has to be filled, even though it is outside of traditional project management duties.

Team Interaction through Technology

It is fundamental to project work that all project teams exchange ideas and project data. This exchange requires information sharing across the project team. By nature, virtual project teams have to depend heavily on information technology to share information through communication and collaboration and to assist them in their day-to-day work activities.[30] Core project management and team leadership skills are of course a greater factor in success or failure of virtual projects, but if we didn't have the information technology tools we have today, virtual projects would not be possible.

The heavy reliance on technology is a key differentiator between traditional and virtual projects, even though technology use on traditional projects has increased. The difference is that the separation of time, distance, and sometimes culture requires technological tools that allow for effective *asynchronous* interaction and collaboration. Additionally, *synchronous* tools, such as audio conferencing technologies, must be chosen properly to improve, not inhibit, effective distributed team interaction. (See the box titled "Technology Must be Useful to Be Used.")

Technology Must Be Useful to Be Used

Mona Harmond, a global program manager for a well-known consumer products company, describes what used to be a common sight in her company's globally distributed offices. "Picture a group of people sitting at a table, surrounding a speakerphone in the middle, trying to conduct a team meeting with three other groups of people in three different geographical locations. If it was a well-run meeting, a common PowerPoint presentation would be projected in each of the three conference rooms and used to direct the conversation of the team."

Though this was a common sight, Harmond agrees that it was not a form of optimal team communication and collaboration. "Even though people were communicating verbally instead of relying on email, this form of communication still presented some challenges," she explains. Among the challenges: the largest group of people typically dominated the conversation; many side discussions took place while other locations were talking; more introverted team members failed to participate in the discussion; and the collective team often failed to fully understand the meeting outcomes, next steps to be taken and by whom, and key decisions that had been made.

"What we learned from these early experiences," explains Harmond, "was that technology, if not selected and used properly, will result in teams reverting to tried and true technologies even though they are not the most optimum solution."

Harmond's learning is consistent with virtual team studies that show that the vast majority of virtual teams rely on email and phone conferences to do their work. It is important to note that while technology will not solve communication problems, it should serve to improve them.

On a day-to-day basis, technological tools enable the generation of work outcomes, online collaboration and review of deliverables, review of tasks completed and tasks to be performed, collection and monitoring of team progress, tracking progress against team metrics, and storing of project information, to name just a few. The primary role of technology in distributed teamwork is one of overcoming the challenges created by time, distance, complexity, and the diversity of participants on a virtual project.[31] Successful use of technology on a virtual project hinges on understanding how the team will communicate and collaborate, understanding how technology can be leveraged to improve team communication and collaboration, matching technology selection to communication and collaboration methods and practices, and then using the technology efficiently to improve the team's performance.

Senior leaders and project managers should select electronic technologies that best meet the needs and usage of the virtual project teams and that integrate with the current suite of tools used within the organization. Note that there is no ideal set of technologies for all teams. A clear

plan for matching technology options to the communication and collaboration needs of their project teams must therefore be developed for a virtual project by the management of the organization and the identified virtual project manager. (See Chapter 8.)

Diligent Monitoring Required

As in all projects, monitoring of tasks and project progress in a virtual project is an essential element of managing to a successful outcome. It should be recognized, however, that managers of traditional projects have a distinct advantage when it comes to monitoring progress on their projects because their teams are co-located. The advantage comes in being able to manage by walking around to gain a sense of the project progress on a daily basis, if desired. This management method gives project managers the ability to verify that what the team is doing is similar to what is being reported. For virtual projects, this verification process is more difficult.

Two things are required to adequately monitor progress on a virtual project: a more formal project reporting system and additional effort on the part of project managers to collect work and project status information.

On a traditional project, project managers can easily call a meeting with little advance notice and preparation to discuss project status on a particular issue facing the team.[32] This, of course, takes advantage of the physical proximity of team members. Managers of virtual projects, however, must be much more organized and proactive because spontaneous or near-spontaneous meetings are rare due to the difficulty in getting the right participants in a meeting on short notice because of the physical and time distribution of the team.

Virtual projects require the institution of a more formal project monitoring and reporting system to keep team members informed of progress of individual tasks and the project as a whole. A formal system defines a specific sequence of reporting, format, and frequency of data input to the system, meeting requirements, and the type, quality, and frequency of progress reports generated.

A formal reporting system does not free virtual project managers from staying in communication contact with team members. On the contrary, virtual project managers must be diligent about continually pulsing team members, albeit by electronic means, to ensure work is progressing as planned and that no issues are blocking team member efforts. See the box titled "When Managing Virtual Teams" to read one project manager's description of how she changes her practices when managing virtual projects.

When Managing Virtual Teams

Shlomit Shteyer, an experienced project manager in the high-tech industry, explains some of the project monitoring nuances she uses when managing a remote team on her virtual projects.

"When working with a local team, the process of reporting is pretty informal. I have a weekly progress meeting, but I'm much more hands-on during each day. I usually tour the office almost every morning and see what's new."

Shteyer goes on to say that "when working with remote teams, progress reporting is much more structured. We set the days we meet, the days I will receive an email status report, and all data is sent to a formal reporting system. My project analyst then generates weekly and monthly project status reports."

The purposeful follow-up and follow-through on behalf of the virtual project manager is critical for team cohesion and project success. "I still need a sense of proof that tells me how the team is doing compared to what they are reporting on. This is where management by walking around becomes so important. For remote teams, I will check in with team members via instant messaging to ask how things are going, if they need my attention on anything, and so on. How often I check in with team members will change based on the level of trust that the information being supplied to me via the reporting system matches what I believe to be the work that is being performed by team members."

Transitioning to the Virtual World

Following the conversation with his manager, Jeremy Bouchard felt both relief and renewed confidence that he could succeed as a virtual project manager for Sensor Dynamics. The relief came from the realization that his manager understood the paradigm shift that Bouchard was adjusting to in moving from traditional project management to virtual project management. Renewed confidence came from hearing that his manager was aware of Bouchard's credentials as an accomplished project manager and of Norville's offer to directly coach him through this transition period.

That being said, Bouchard also realized that he has a lot of work and learning ahead of him as his journey from traditional to virtual project management has begun. Following the advice of his manager, Bouchard will focus on three primary perspectives: (1) what aspects of traditional project management stay the same for a virtual project; (2) what aspects have to be modified to be effective on a virtual project; and (3) what new project management methods, tools, and practices have to be learned and adopted for use on a virtual project.

Assessing the Virtual Project Manager

The Virtual Project Manager Assessment measures experience and competency. It is used and is useful for matching available virtual project managers to virtual projects based on each project leader's skills and experience. The results give management the opportunity to evaluate the expected difficulty, complexity, and uniqueness of each of the virtual projects to the right mix of skills and experience of the available project managers.

The assessment can also be useful in identifying training and development needs for the organization's virtual project managers and for the recruiting and hiring of new ones.

It is recommended that either one or several managers complete the assessment tool. If more than one manager completes the tool, they should discuss responses to each item and determine how best to align virtual project manager resources to the virtual projects being planned.

Virtual Project Manager Assessment

Date of Assessment: _____

Virtual Project Manager Name: _____

Assessment Completed by: _____

Confidential Assessment: _____ Yes, confidential

_____ No, not confidential

Assessment Item	Yes or No	Notes for All "No" Responses
Project Management Experience		
Has experience managing virtual projects.		
Has over five years of experience successfully managing traditional projects.		
Has the proven ability to manage cross-team deliverables and integration across multiple project sites.		
Has the proven ability to communicate effectively using electronic tools.		
Has demonstrated the ability to apply project management skills successfully in a virtual project setting.		

Assessment Item	Yes or No	Notes for All "No" Responses
Team Leadership Experience		
Is goal-oriented, self-directed, and motivated.		
Has the proven the ability to create and achieve a common purpose.		
Has proven to be accountable and meet commitments.		
Has demonstrated a high degree of personal integrity.		
Has the proven ability to leverage emotional intelligence skills in order to manage situations, people, and deliver business value.		
Has the proven ability to leverage contextual management skills in order to manage situations, people, and deliver business value.		
Has the proven ability to negotiate effectively.		
Has the proven ability to build positive working relationships with distributed stakeholders.		
Has the proven ability to efficiently and effectively manage conflict between virtual team members.		
Has the proven ability to make tough decisions.		
Has the proven ability to resolve problems, remove barriers, and accomplish goals while leading a distributed team.		
Has the proven ability to create shared values and provide recognition selflessly.		
Exhibits confidence and is well respected.		
Virtual Team Skills and Experience		
Has the proven ability to establish team chemistry virtually.		
Has demonstrated the ability to utilize systems thinking skills effectively in the virtual environment.		
Possesses political acumen necessary to influence company leaders virtually.		
Has the proven ability to drive virtual participation and collaboration.		
Has the proven ability to create and manage cross-cultural awareness.		

Assessment Item	Yes or No	Notes for All "No" Responses
Has the proven ability to leverage technological tools to facilitate team communication and collaboration.		
Has the proven ability to build and sustain trust between distributed members of the team.		
Has the proven ability to facilitate effectively across multiple project sites.		
Has the proven ability to network effectively.		
Has the proven ability to empower distributed team members.		
Has the proven ability to select and manage virtual team members.		
Has the proven ability to delegate tasks properly to a virtual team.		
Business/Financial Skills		
Has demonstrated strategic thinking skills.		
Has demonstrated the ability to align project goals to business goals and strategies.		
Has demonstrated the ability to apply business, financial, and cost fundamentals to a project.		
Has the proven ability to apply worldview skills in context to the nations involved with the project.		
Customer/Client Skills		
Has the proven ability to learn customer and client needs and convey those needs effectively to others on the team.		
Has the proven ability to meet customer and client demands.		
Has the proven ability to gain customer commitment.		
Has the proven ability to achieve customer quality expectations.		

Findings, Key Thoughts, and Recommendations

Notes

1. Global Workplace Analytics, www.GlobalWorkplaceAnalytics.com.

2. R. Lepsinger and Darleen DeRosa, *Virtual Team Success: A Practical Guide for Working and Leading from a Distance* (Hoboken, NJ: John Wiley & Sons, 2010).

3. Jim Waddell, Tim Rahschulte, and Russ J. Martinelli, "Leading Global Project Teams," *PM World Today* 12, no. 7 (July 2010).

4. Trina Hoefling, *Working Virtually: Managing People for Successful Virtual Teams and Organizations* (Herndon, VA: Stylus, 2003).

5. "Managing Virtual Teams: Taking a More Strategic Approach" (London, UK: Economist Group, 2009). www.economistinsights.com/business-strategy/analysis/managing-virtual-teams.

6. Waddell et al., "Leading Global Project Teams."

7. Adam Smith, *An Inquiry into the Nature and Causes of the Wealth of Nations* (New York, NY: Modern Library, 1776/1994).

8. Roger Bootle, "We Now Face Keynesian Conditions and Need Truly Keynesian Solutions," *London Telegraph*, October 26, 2008. http://www.telegraph.co.uk/finance/comment/rogerbootle/3264845/We-now-face-Keynesian-conditions-and-need-truly-Keynesian-solutions.html.

9. Waddell et al., "Leading Global Project Teams."

10. Manfred B. Steger, *Globalization: A Very Short Introduction* (London, UK: Oxford University Press, 2013).

11. Waddell et al., "Leading Global Project Teams."

12. Jim Waddell, Tim Rahschulte, and Russ J. Martinelli. "Putting Skin in the Game." *PM World Today, 12, no. 9, September 2010.*

13. Waddell et al., "Putting Skin in the Game."

14. Steger, *Globalization.*

15. Waddell et al., "Leading Global Project Teams."

16. Waddell et al., "Putting Skin in the Game."

17. Thomas L. Friedman, *The World Is Flat* (New York, NY: Farrar, Straus and Giroux, 2006).

18. Hoefling, *Working Virtually.*

19. Mark A. Filippell, *Mergers and Acquisitions Playbook: Lessons from the Middle-Market Trenches* (Hoboken, NJ: John Wiley & Sons, 2010).

20. Hoefling, *Working Virtually.*

21. Russ J. Martinelli, James Waddell, and Tim Rahschulte, *Program Management for Improved Business Results*, 2nd ed. (Hoboken, NJ: John Wiley & Sons, 2014).

22. Waddell et al., "Leading Global Project Teams."

23. Waddell et al., "Leading Global Project Teams."

24. Terri R. Kurtzberg, *Virtual Teams: Mastering Communication and Collaboration in the Digital Age* (Santa Barbara, CA: Praeger, 2014).

25. Parviz F. Rad and Ginger Levin, *Achieving Project Management Success Using Virtual Teams* (Plantation, FL: J. Ross, 2003).

26. Rad and Levin, *Achieving Project Management Success.*

27. Rad and Levin, *Achieving Project Management Success.*

28. Martinelli et al., *Program Management for Improved Business Results.*

29. Rad and Levin, *Achieving Project Management Success.*

30. R. Jones, Robert Oyung, and Lise Pace, *Working Virtually: Challenges of Virtual Teams* (Hershey, PA: CyberTech, 2005).

31. D. Duarte and Nancy Tennant Snyder, *Mastering Virtual Teams: Strategies, Tools, and Techniques that Succeed* (San Francisco, CA: Jossey-Bass, 2001).

32. Rad and Levin, *Achieving Project Management Success.*

PART

II

PLANNING THE VIRTUAL PROJECT, BUILDING THE VIRTUAL TEAM

2

PLANNING THE VIRTUAL PROJECT

There was a lot of excitement at this month's portfolio review meeting at Sensor Dynamics. The company's researchers achieved a technological breakthrough that provides a pathway into the new autonomous automobile market. It was believed that the technology was ready to go into a product, and strong interest had been expressed by both new players in the auto industry, and a few large technology companies, and historic automobile manufacturers. Sensor Dynamics received a commitment from a Japanese automobile manufacturer to sign on as a strategic partner. The portfolio team at Sensor Dynamics approved the development of the new sensor product, code named "Sitka," and added it to the company's new product development portfolio.

Jude Ames, Vice President of New Product Development, turned to Brent Norville and asked an important question: "Do you have a PM and team that you can put on this project now?" The obvious answer was no because every project manager and team were currently committed to projects, but Norville knew this answer would not be acceptable. "Everyone is committed to a project right now," began Norville, "but Jeremy Bouchard's team will be launching their biometric product in three weeks, so we can probably move the entire team over to the Sitka project. In fact, Jeremy and a few of the project leads could begin early initiation and planning work next week."

As all good senior executives know how to do, Ames asked another important question: "Is that the right team to put on this project? Success is critical." Norville was aware that Bouchard had

some challenges moving into a role of managing virtual projects, but much of that had to do with the situation in which he was put. In particular, his problems had arisen when he was asked to manage a project and team that was highly distributed and a team that had not worked together in the past. By keeping the team together for the Sitka project, Bouchard and his people would be in a much better position this time to begin performing immediately. "By keeping the virtual project team mostly intact, Jeremy should be able to get the Sitka team formed quickly and get them focused on project planning," Norville stated. "The challenge will be bringing the Japanese auto partner into the team and integrating the new technologists from our research facility in Europe," he added.

Bouchard welcomed the opportunity to begin initiating the Sitka project, particularly since much of the Sitka project team would consist of people he has had on his team for the past 18 months. "On my first virtual project, I was overwhelmed during the planning phase because I really didn't understand the amount of effort that is required to form a virtual project team," explains Bouchard. "It's truly like having two full-time jobs—planning the project and forming the project team." Like Bouchard, many project managers experience the shift in the amount of effort required to fulfill this dual role when they move from managing traditional projects to managing virtual projects.

Even though the terms *project manager* and *project leader* are used interchangeably, it is important for project managers to understand that the work associated with the two roles is *not* interchangeable. To help delineate between the two roles, remembering the distinction described by author John Maxwell is helpful: "Managers work with processes and leaders work with people."[1] Obviously, project managers do both. As Jeremy Bouchard reflected, this is especially true and important during early initiating and planning of a virtual project. Knowing where to focus your efforts in each of the two roles in order to gain maximum value for those efforts is the subject of this chapter and the next one.

Planning a Virtual Project

Chapter 1 described the primary differences between virtual and traditional projects that have to be considered specifically when managing a virtual project. These differences do not take long to emerge and for their effects to be felt. Most, if not all, emerge when a project is in the earliest stages of the project life cycle, as described in the next box.

A Project's Life Cycle

To manage a project effectively, structure is needed to guide the team's work. A typical structure is the project life cycle—initiating, planning, executing, and closing. As noted in PMI's *Guide to the Project Management Body of Knowledge*, "A project life cycle is the series of phases that a project passes through from its initiation to its closure."[2]

There is significant debate regarding the most appropriate life cycle model that project managers should use to lead work. Harold Kerzner and others have spent a significant amount of time writing books and articles detailing life cycles used by project teams in industries including engineering, manufacturing, software development, and construction. Indeed, a number of models have been established over the years. Some are superficial and high level, while others are elaborate and intricately detailed. Some are industry specific, while others are project specific. In short, there are a number of project life cycle models available for use.

Because of Kerzner's fine contributions such as his book entitled, ***Project Management 2.0***, there is no need to replicate that work here in detail. If we take a step back for a moment from these models themselves, however, we can realize the core constructs of projects themselves. It is easily agreed that projects vary in size, scope, and complexity, but regardless of such differences, all projects follow a rather similar, albeit generalized, life cycle from start to finish—from ideation to project closure. (See Figure 2.1.)

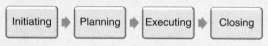

Figure 2.1: Basic Project Life Cycle Model

Regardless of whether the life cycle used is predictive, iterative, incremental, or adaptive, the phases generally are sequential and are broken down by objectives, deliverables, milestones, scope of work, or investment allocation. Importantly, each phase is time bound, which emphasizes a control point for each phase (or investment). The detail of any life cycle (number of phases and description of each phase) is often determined by organizational culture, risk tolerance, and project discipline maturity. The project life cycle serves as the framework for managing the project. Thus, an appropriate level of structure must be applied. Too much structure can bog the project team's effort down in bureaucracy. Too little structure, and the team can drift outside of scope and miss business goals.

Most individuals who have been part of a virtual project know the challenges very well. Some, however, may be wondering what all the fuss is about and why virtual projects are so much more difficult than traditional projects. To understand the challenge, let's use an example in which effective communication is of paramount value, as is the case with any project, large or small, traditional or virtual. Think of the game commonly referred to as the Telephone Game. You may recognize this game by one of its other names, such as Chinese Whispers, Russian Scandal, and Secret Message. The game goes like this: One person whispers a message or story to another person, who in turn passes the message (or at least what was heard as the message) to another person, who in turn passes the message to another person, and so on until the point when everyone in the game has heard the message—or, perhaps more specifically, heard *a* message. To conclude the game, the last person says to first, or to the entire group, what he or she heard as the message.

As can be expected, errors in the message accumulate as the interpreting and retelling of the story occurs from one person to another. The statement announced by the last player in the Telephone Game usually differs greatly from the initial message. This is because of listening and transferring errors, translation and comprehension issues, and poor communication from which inferences and assumptions are added and details are misrepresented or omitted. All of this misinterpretation and misunderstanding stems from the challenge of communicating effectively.

Interestingly, these are all problems that occur when players in this game are in the same room or physical space. Now imagine this game getting played out over a virtual space, one phone conversation or one email after another. The same challenges and issues not only occur, but they are magnified because of time lapse as messages cross time zones, cultural and language barriers due to first-language preference and culture-specific values, and other filters often used to interpret, translate, and respond.

This is only one example of the many challenges associated with managing virtual projects. However, it demonstrates a subtle but very important factor that all virtual project managers need to know. The fundamentals of managing a virtual project are the same as the fundamentals of managing a traditional project. But there are important nuances concerning how the project management fundamentals are practiced on a virtual project due to the differences

that were detailed in Chapter 1. These nuances have a number of important implications when it comes to managing a virtual project and leading a virtual project team. First, they elevate some of the core project management fundamentals in importance over others. Second, how the fundamentals are practiced and the processes are used need to be modified for use in the virtual project environment. Third, simplicity is better. Finding ways to practice the project management fundamentals in the simplest manner helps to reduce virtual project complexity instead of adding to it.

A Few Words about the Term *Nuance*

We use the word *nuance* to describe the differences between managing virtual and traditional projects. Now, *nuance* is commonly defined as a subtle difference.[3] We want to emphasize that while the differences may be subtle or small, there are a lot of them, and when they are aggregated, the collective impact of each nuance creates a multiplying effect and challenge for project managers. All the nuances together serve to magnify the challenge. So, we don't mean to minimize the differences in virtual and traditional projects with the use of the word *nuance*.

Virtual Project Planning Process

One of the key differences between traditional and virtual projects discussed in Chapter 1 is the importance of purposeful integration of distributed work outcomes. Integration takes center stage during project planning as it is the responsibility of the project manager to ensure that an integrated, coherent project plan is developed despite the fact that most of the plan details and elements are first created by project team members who may be widely distributed both organizationally and geographically. Even though a virtual project team may be highly distributed, integration of work must be centralized around the project manager.

In Chapter 6, we discuss at length how a centralized/decentralized model is used to perform much of the work on a virtual project, but we introduce it

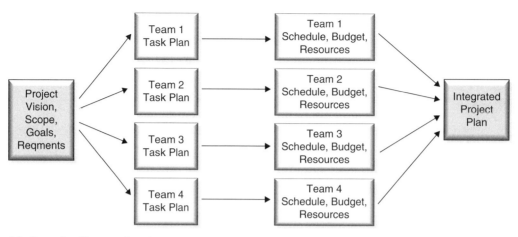

Figure 2.2: Centralized/Decentralized Project Planning Model

here because the model is central to the planning process for a virtual project. Figure 2.2 provides an illustration of the centralized/decentralized planning process.

Like traditional projects, the critical information needed to plan a virtual project is first centralized and created by the project manager and his or her project planning team. This information consists of the project vision, project goals and success criteria, project scope, and the detailed project requirements. While this is the minimum information needed, experienced virtual project managers know that the following must be established to offer the greatest probability of virtual project success:

- Focus on deliverables before tasks
- Establish distributed task plans
- Integrate the project plan
- Gain project plan approval

Deliverables before Tasks

With the project scope defined and documented, the natural tendency is to move into detailed task planning. However, prior to this, many seasoned virtual project managers first go through the step of mapping project deliverables over a timeline. Co-located teams tend to manage their work by

completion of tasks, whereas distributed teams more effectively (and visibly) manage their work by completion of deliverables.

A process called project mapping is often used to identify the critical deliverables throughout the life cycle for each team on a project. More important, the outcome of the project mapping process, the project map, shows the cross-team interdependencies that exist between the members of the team.[4] Virtual project team members are expected to act as more than just individual experts pursuing their own specific tasks; they are team members focused on tasks and the cross-team interdependencies needed to effectively deliver an integrated solution. An explicit recognition of the various mutual dependencies between the distributed members of the team is needed, especially in the virtual project environment.

Figure 2.3 illustrates a partial project map that shows the deliverables and cross-team interdependencies during a two-month period of the project. Each deliverable identified in the project work breakdown structure is displayed on the project map in the correct sequence and relative point in time. Arrows between deliverables depict the interdependencies among project members (what deliverables are needed by whom and when).

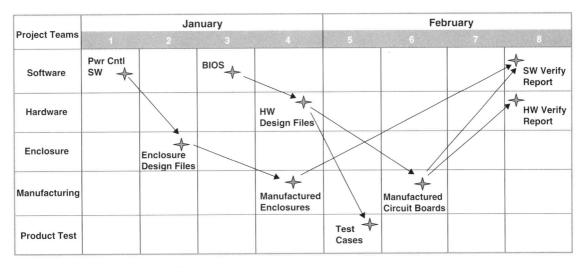

Figure 2.3: Example Partial Project Map

The mapping of deliverables among project team members helps the team determine and fully appreciate the dependencies that exist on a project.

By following one of the deliverable chains, we can demonstrate how cross-team interdependency and collaboration is organized in this example. Pwr Cntl SW (power control software) is a deliverable generated by the software development team and delivered to the enclosure development team. The enclosure development team in turn uses the power control SW deliverable to generate its enclosure design files deliverable, which is handed to the manufacturing team. The manufacturing team uses the enclosure design files deliverable to generate its next deliverable, the manufactured enclosures, which are then delivered to the software development team, which uses the deliverable to generate its SW verify report. This mapping of deliverables shows the critical interdependencies among the software, enclosure, and manufacturing teams on the project.[5]

Virtual project managers should lead their teams through the creation of the project map early in the planning phase and then utilize it as a tool throughout the remainder of the project life cycle to drive collaboration, synchronization, and integration of work output across the distributed team. More on this in Chapter 4.

Distributed Task Plans

With the project-level planning information and a map of project deliverables and interdependencies in hand, detailed planning activities can be delegated and distributed to the members of the virtual team. Detailed planning is distributed to the locations where the work will be performed. Each person responsible for detailed planning will provide task plans for developing deliverables, a timeline to do so, a resource plan, budget required, base assumptions, and any identified risk in completing work and the quality of the work outputs.

To keep the work of the various detailed planners synchronized, project managers need to set clear milestone dates for detailed plan creation and establish periodic reviews to ensure work is progressing. Should any changes to project scope, requirements, or goals occur, project managers must quickly communicate the changes to the entire team so the changes can be incorporated in the detailed plan. Once the detailed plans are completed, project managers again centralize the project planning process by leading the project team through the creation of the integrated project plan.

Integrated Project Plan

The integrated plan incorporates the detailed plans of the distributed team members and presents that work as a coherent set of activities to be performed. Although the process for developing an integrated project plan differs from that used for a traditional project, the plan itself will be traditional in content. (See Table 2.1.)

Project Plan Approval

As explained in Chapter 1, decision making on a virtual project can be considerably slower than on traditional projects. Decisions necessary to formally approve the integrated project plan are particularly slow to make. This is due in large part to the organizational and geographic distribution of key stakeholders who serve as project decision makers.

Virtual project managers do not have the luxury that their traditional project counterparts have of getting the decision makers together and driving to a final approval decision. Rather, multiple meetings and multiple electronic communication exchanges

Table 2.1: Project Plan Checklist
✓ Project purpose
✓ Work breakdown structure / scope
✓ Project objectives
✓ Key deliverables and project output
✓ Success criteria and measurement metrics
✓ Detailed schedule
✓ Detailed budget
✓ Resource profile
✓ Risk analysis
✓ Tracking and reporting methods
✓ Communication plan
✓ Change management plan

with multiple stakeholders will likely be required to converge on a final decision.

To help alleviate project plan approval decision delays, virtual project managers can do a number of things. First, they can ensure that there is a single approver for the project plan and that all project stakeholders are keenly aware of who that person is. Often decision delays on virtual projects are caused by lengthy debates among the various stakeholders about the identity of the decision maker. Of course, often many stakeholders believe *they* should own the decision rights. It is important to get clarity *before* a decision is needed and to document it. Once the decision maker is identified, documented, and communicated, virtual project managers can take the decision stakeholder identification process one step further and identify who will provide input to the decision maker concerning approval of the project plan. Again, many stakeholders may slow the decision process by behaving as if they have the responsibility and authority to guide the decision maker on what his or her decision should be. Some project stakeholders probably are in this position; others will not be. Virtual project managers must document who has decision input authority and make sure it is clearly communicated among the team and stakeholders.

Another cause of decision delay is the lack of necessary data for decision makers to actually make the decision to approve the project plan, or not. To speed this decision, virtual project managers should be aware of the data needed for the decision; that data collection must become a primary focus of the planning process.

Finally, virtual project managers must make sure that they, as project managers, provide a recommendation for project plan approval (or not, as the case may be) and justify their recommendation. Project managers should know as much about the viability of the project and the likelihood of project execution success as anyone else in the organization at the time of project plan approval. Virtual project managers should respectfully state their opinion, and provide justification as needed, to their senior management.

Scope, schedule, cost, and quality serve as the foundational underpinnings of success on any project, traditional or virtual. The fundamental practices and processes for establishing the project scope, developing a realistic project schedule, creating the project budget, and documenting the quality requirements during the planning stage of a virtual project are identical to those used for traditional projects. However, three challenges emerge during project execution that requires special attention outside of creating the integrated plan during project planning. These challenges are:

1. It is more difficult to maintain alignment between project outcomes and business strategy on a virtual project and to maintain alignment among project stakeholders who are commonly geographically distributed.

2. The virtual project structure cannot form organically as is the case for many traditional project teams. Rather, it has to be purposefully architected.

3. Virtual projects are more complex. Complexities must be made visible and associated risks identified to simplify the complexity of virtual projects.

These challenges are due primarily to the physical distribution of stakeholders and the project team. How virtual project managers can mitigate each of these challenges during the planning stage of the project is discussed in the sections that follow.

Establishing Project Alignment

Even with an integrated project plan in place that has been approved and communicated, the distributed nature of the virtual project environment applies a constant force that can fragment the plan during project execution and cause major misalignment. This misalignment occurs among project team members, between the project team and stakeholders, and between what the project ultimately delivers and the anticipated business goals. To battle these forces, paying special attention to the establishment of project alignment is a critical virtual project planning practice.

Alignment is often defined as a straight-line arrangement of things or as an alliance among people. When thinking about alignment relative to individual work, it means each person (on the team) can understand their work and its meaning relative to the whole of what is being created by the team as well as the importance of their collective work in achieving the business goals of the specific project and enterprise strategy.

Alignment does not happen by chance. Rather, alignment is created (and sustained) by intentional influence. That means it is planned. Purposeful alignment is necessary on virtual projects, especially on projects that have little or no co-location among team members. The primary reason for this is that team members have limited opportunities to engage in informal discussions about why they are doing a particular project and to see how the work of each individual is integrated to the larger, whole solution. For these reasons, alignment has to be specifically and intentionally planned, documented, and communicated in ways that are clear and will be remembered.

To ensure alignment between a firm's strategy and its execution outcome, the senior leaders of an organization, the project managers, and various project team members must share a common understanding of what the project has been commissioned to achieve, what will be delivered as the project outcome, and how the outcome will be developed. Unlike traditional projects, where the critical players have direct access to one another, developing this common alignment of thinking is more difficult on a virtual project. To overcome the barrier created by distributed organizational stakeholders, project managers must work diligently in the early stages of a project to align team members to the solution to be created, with an understanding of how their efforts will contribute to the whole solution. Doing so creates a common project vision. Alignment also involves ensuring that the project supports the accomplishment of the underlying business goals and strategies. Finally, project alignment involves defining team success and the ground rules of how the team will collaborate and self-govern.

Aligning to the Vision

All project teams need a vision of what is being developed through the labor of the collective team's work. The reason for this is rather simple. It is best to have a vision of the final outcome as you work on individual parts of the whole. It is for this reason that architects build small-scale models of an envisioned skyscraper. For similar reasons, auto designers create sketches and manufacturers build prototypes. If you cannot see the whole, it is difficult to understand the value of individual parts.

In *The Fifth Discipline*, the seminal book on systems thinking, Peter Senge details the need to see the "whole" of things.[6] He mentions that one of the greatest challenges individuals and teams face

is the inability to see the whole of our work and the impact from that work. It is for these reasons that when we are working with project teams to establish a common vision, we use the term *whole solution* to describe the final outcome—the product, service, infrastructure solution, or other capability that the team will create and introduce into the market or organization (see the box titled "Understanding the Whole Solution Concept"). Figure 2.4 illustrates a simple example of a whole solution.

In this example, the development of a smartphone, the whole solution consists of all of the elements necessary to create the total solution for the customers of the company. Obviously, it consists of the various hardware, software, and wireless communication elements of the physical product, but it also consists of ancillary elements, such as packaging, manufacturing, infrastructure enablement, marketing, and customer support. If any one of the primary elements of the whole solution is missing, the product would fail to meet customer needs.

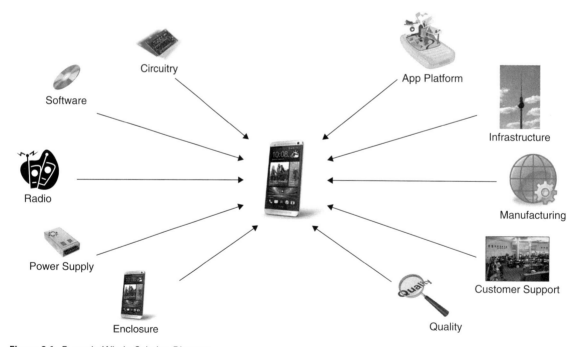

Figure 2.4: Example Whole Solution Diagram

In a virtual team environment, the development of the elements most likely would be distributed to various team members and partner organizations, potentially across the globe. By creating and communicating the whole solution, each member of the virtual project team begins to see the holistic view of what the team is creating and how their work output contributes to the project mission and vision.

When completed effectively, the whole solution shows that success cannot be fulfilled by any one specialist or set of specialists on the team. Rather, success comes when meeting customer expectations is a shared responsibility among the members of the project teams, with their work tightly interwoven and driven toward the integrated solution even though project members are distributed across geographic boundaries.[7]

It can be argued that there is no point in the project life cycle more important to see the whole solution than in the initiating and planning phases of project work. This is a formidable time for project team members and key stakeholders because, on one hand, the project (really) hasn't started yet since full funding to execute has not occurred; but on the other hand, this is the time the team is being formed, expectations are being set, and perceptions of individual and collective work are being established. We have all heard the phrases "Get off to a fast start," "Plan for early wins," and "Build early momentum." The value of a fast start, in reality, is more than just momentum. It also includes structure, buy-in, purpose, and meaning, all of which come from alignment of people to goals. This alignment is achieved when project managers can establish purpose and meaning of work with connection between members of the virtual team relative to the whole solution.

Understanding the Whole Solution Concept

The concept of the whole solution is not new. It originated when Geoffrey A. Moore coined the phrase the "whole product" in his book *Crossing the Chasm*.[8]

Moore used the phrase to describe aspects of marketing organizational value. There are two compelling value propositions that are quite important to understand this concept:

1. The expectation on the part of customers that their wants and needs be met.

2. The ability of the company to provide an *integrated* capability that fulfills the wants and needs of customers.

Many times there is a gap between the two. To close the gap, the company must add an array of services and ancillary products to the original solution, thereby creating the *whole solution*. Therefore, the whole solution is the product that provides the *maximum* chance to achieve customers' buying objectives.

Aligning to Business Goals and Strategy

To provide the greatest value to a company, there must be direct alignment between the outcome of a project and achievement of one or more business goals of the company. Experienced project managers know that there is a direct correlation between project duration and alignment risk. The longer the project's timeline, the more inherent risk there is that alignment to strategy and goals may shift as the project team attempts to accommodate changes in markets served, product requirements, competitor actions, or customer demands. This is why it is so important to clearly align the strategy and goals among senior leaders, the project team, and other stakeholders during project planning. The project business case is a critical tool used to help establish and sustain alignment throughout the project (see the box titled "The Project Business Case"). To be effective in setting alignment for a virtual project, the business case must spell out the business benefits and the rationale as to why the proposed outcome is desired by customers, users, and the organization. This information is critical for all projects but is essential for virtual projects. It answers the question, "Why are we doing the project?" that creeps into the thinking of most team members (and many stakeholders) at one time or another. For team members who work in remote locations and cannot engage in informal discussions about the underlying reason and purpose of their project, the business case is their foundation for remaining aligned to the rest of the team and the organization as a whole.

The Project Business Case

In the book *Project Management ToolBox*, authors Russ Martinelli and Dragan Milosevic detail the function and use of the business case.[9] The business case is used to:

- Gain agreement on project scope and success criteria.
- Gain approval for resource allocation and funding.
- Evaluate the project relative to alternative projects in the portfolio.

When starting the business case in the early stages of a project, it is known that not all data will be available. Therefore, assumptions will need to be made regarding the business environment, customer requirements, business strategy alignment, and the business success criteria for the project under consideration. These assumptions will need to be validated during detailed planning.

In addition to clarifying the project's purpose, scope, timeline, and cost, the business case should answer these questions:

- How much will this project contribute to the company's bottom line?
- Is the outcome worth the investment?
- What's the probability of success for this project?
- What will be done to maximize the probability of success?
- How will the known risks be avoided or mitigated?
- Does the level of risk prevent investing in the project?

Experienced project managers know that the business case is the culmination of information gathered in the initiating phase of the project and is used to establish validity and the means by which the project will prove to be a successful investment in the enterprise portfolio. The business case includes a problem statement and outlines the business value by way of opportunity analysis. This analysis assists in determining the proper fit of the project under consideration relative to other projects based on funding available, strategy, market positioning, and risk. Furthermore, a feasibility portion of the business case summarizes the financial net benefit to the organization if the project is approved. Table 2.2 describes the elements of the project business case.[10]

Aligning the Team

All project charters should provide information about the expected result from the execution of the project. The best project managers know that the project charter needs to balance high-level strategy with lower-level detail.

At the strategy level, the charter needs to describe how the project team will achieve business goals and deliver value. The charter establishes this by defining the project objectives and the capability outcome from the project (the new product, service, infrastructure, or other capability that will result because of the project outcome). Importantly, in the charter, the project objectives must align to the business strategies of the organization and

Table 2.2: Minimum Elements of a Project Business Case	
Business Case Element	**Description**
Purpose	A succinct statement of the business benefits driving the need for the project investment.
Value proposition	A succinct statement characterizing the value to be delivered (quantified when possible).
Benefits identification	A comprehensive list of the important benefits to be derived by completion of the project. The result of this effort is what the Project Management Institute refers to as the Benefits Realization Plan.[11]
Risk analysis	A comprehensive analysis of the critical risks to the project and the planned mitigation efforts to address each risk.
Business success factors	The set of quantifiable measures that describe business success for the project.
Governance model	The project's oversight team and process for monitoring project progress.
Detailed cost analysis	The investment cost of the project and timing of the costs.
Critical assumptions	The events and circumstances that are expected to occur for successful realization of the objectives.
Project timeline	A project schedule identifying the critical milestones, decision points, and timing expectations by senior management and other key stakeholders.

provide the guiding principles that help the project manager and other key decision makers understand the scope of the project and determine resource requirements necessary for the project's success. To be effective, objectives should be specific, measurable, achievable, realistic, and time bound.[12]

Project Charter

There are a lot of commonalities between project charters and project scope documents. The factor that is different is *detail*. The charter is often used as a decision making tool, whereas the scope document is used as a planning tool. Therefore, the charter often includes less detail because it is used to present to executive teams. The scope document is more detailed and used by the core project team to plan project work.

Team Charter

In addition to a project charter, many virtual project managers also create a team charter during project planning activities. The team charter can be appended to the project charter or be a separate and stand-alone document for the project team's use.

As noted, one of the most common challenges for project managers responsible for a virtual team is alignment. Most project managers responsible for traditional projects follow advice from PMI in their Project Management Body of Knowledge publication and, in doing so, miss this important nuance. Although the project managers clearly detail the project's charter, they miss the intentional beginnings of creating a team charter.

Ideally, it is created in a group setting, in realtime, and details direction and boundaries, processes and protocols to resolve conflict, and expectations and practices that govern individual work in a way that, over time, builds trust, respect, and rapport. You can think of the team charter as a governance-type tool that details how work will get done among the project team. It identifies ground rules, expectations, and values. Table 2.3 highlights the key elements of a team charter.

The team charter is a critical tool for virtual project teams. Further, the team charter helps to detail elements of a highly effective team; it essentially is a plan on how to get the virtual project team right. Keith Ferrazzi identified four absolutes when it

Table 2.3: Key Elements of a Team Charter

Team Charter Element	Description
Team name	Just like products have names, teams should have a name. Names unify team participants.
Team goal	It's best to have the team define its goal. The project charter and scope document will detail a goal, but the purpose here is to have the team collectively describe it, write it in a (short) narrative form, and illustrate it if possible.
Team success metrics	Project and business metrics will be outlined in the project charter and scope document, but they may not include specific team success metrics. Team success metrics provide the team an opportunity to detail how each person will define success relative to another.
Conflict resolution	Understanding how team conflict will be resolved is important. It is even more important to acknowledge that conflict likely will happen. Spending time ahead of any conflict discussing how to resolve conflict is much better than doing so at the time of the first conflict.
Unconscious bias	Virtual team members often have biases based on myth and personal assumptions. This is especially true across cultural barriers. Spending time to demystify myths breaks down barriers and speeds the norming phase of teams.
Value statement	Teams can accomplish more than any individual alone. With the preceding items detailed, the team can detail a value statement. The best team value statements contain the contribution expectation of each individual and the team at large. This statement is a pledge or commitment to one another.
Team check-ins	Like detailing a process for conflict resolution, it is good to chart out a timeline to check on the health of the team before getting far into project execution. The checkpoints should be both one-on-one with the team member and project manager and in group form.

comes to highly effective virtual teams (see the box titled "Planning For The Right Virtual Team").

Planning for the Right Virtual Team

Keith Ferrazzi, CEO of Greenlight, a research-based consulting company, wrote recently about how to get virtual teams right.[13] He raised the question that so many are trying to answer: How do you create and lead an effective virtual team? There is no shortage of advice and opinion in answering this question. Ferrazzi concluded his research with four must-haves for effective virtual project teams:

1. The right team
2. The right leadership
3. The right touchpoints
4. The right technology

Architecting the Virtual Project

Because work on a virtual project can be highly distributed, the project itself cannot form organically in the same manner as traditional projects. Instead, it has to be purposefully architected and structured to ensure all work is defined and distributed systematically and intelligently.

The term *architecture* refers to the conceptual structure and logical organization of a system. It includes the elements of the system and the relationships between them.[14] A *project architecture* is therefore the conceptual structure and logical organization of the project. It is comprised of core project components as well as the noncore functional components required to create and deliver the whole solution.

To create an effective project architecture, it is helpful to again take a systems approach. In any case where a new capability will be created, the project serves as the delivery mechanism for that capability and should be designed and structured in a systematic fashion. The whole solution will dictate the primary project components and enabling components needed for the project architecture.[15] Take, for example, the simplified whole solution diagram for the creation of a new capability such as a smartphone as illustrated in Figure 2.4. The architecture is easily derived directly from the whole solution.

Figure 2.5 demonstrates how the architecture might be designed for the smartphone described previously.

When project teams are complex, as is the case most of the time and as illustrated in Figure 2.5, seeing the team in an architectural diagram can create better, more intentionally designed, and methodically distributed work geographically. It enables the team to see their workgroup relative to others on the team. Additionally, and importantly, seeing the architecture illustrated reinforces the fact that a virtual project cannot form organically. The software and circuitry teams may be able to come together naturally in Palo Alto, but they would not come together naturally with the power supply team in Dallas or the radio team in Sal Paulo. Further, there is no way that the manufacturing team in Shanghai will organically communicate key product

attributes to the customer support team in Sydney. Again, the figure illustrates the intentionality needed to form the project team and carry out project work. If the project team cannot see their work and the work of others, the virtual project manager must illustrate the work so that it can be understood.

Understanding the Complexity of Virtual Projects

All projects have complexity. Designs have become more complex as features and integrated capabilities increase; the process to develop and manufacture the solutions requires more partners, suppliers, and others throughout the value chain; the ability to integrate multiple technologies with end user wants requires not only accuracy regarding requirements delivery but also speed and agility to change; and the current global, highly distributed business environment requires work to occur in multiple sites across multiple time zones. (See the box titled "Managing across Time Zones.") Therefore, the ability to characterize and profile the degree of complexity associated with a virtual project has become essential for both executive leaders and their project managers.

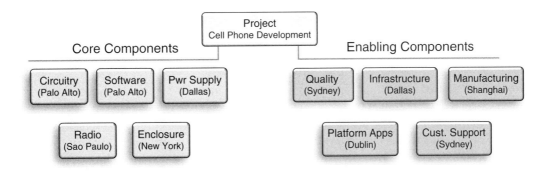

Figure 2.5: Example Project Architecture

Managing across Time Zones

Early ideals of globalization highlighted the value of 24-hour-per-day work cycles. The thought was that products, as an example, could be originally designed in North America, developed in the Far East, tested in the Middle East, modified in Western Europe, and then iterated again in North America. Theoretically, this made for a 24-hour seamless work cycle. However, little is actually realized from such theoretical cycles. A 24-hour work cycle does not equate into 24 hours of productive work. The once-idealized notion of "work while we sleep" is realized most often as "wait while we sleep" because of the lack of smooth transitions between work teams around the globe trying to take advantage of time.

The world is divided into 24 standard time zones with 40 fractional time zones.[16] Global teams experience 12-hour variances in business operations, thus ensuring early-morning and late-evening team meetings and gatherings. It is easy to work across time zone variances in your own country. For example, when colleagues work together in North America, an early-morning meeting is still morning in each of the locations. The time difference between New York and Los Angeles is three hours. So, an 8:00 AM meeting for a team in Los Angeles is an 11:00 AM meeting for the team in New York.

Including team members from India, however, becomes a greater challenge. Consider the time difference between New York and Bangalore. One location will be in the morning and the other in the evening. Another challenge arises because these locations do not have full hour variances, but half hour variances. The difference in time between New York and Bangalore is 10.5 hours. Thus, an 11:00 AM meeting in New York is 9:30 PM in Bangalore. These two locations can squeeze a meeting together in roughly the same workday if the meeting starts at 7:00 AM New York time, which would be 5:30 PM in Bangalore. This meeting would be problematic, however, if we add the team from Los Angeles—where it is 4:00 AM on the West Coast of North America. Such is the challenge currently facing virtual project managers.

Another repercussion of virtual teams is perceived in-group favoritism due to convenience. A member from our research sample confessed, "When you have global responsibility, you tend to spend your time in the regions that may not necessarily need help, but rather that are the most convenient. I spend more time with Europe than I do the other regions because it is easier to get up early and check in with them. The Pacific Rim comes online in the evening, and it is difficult to have a personal life and do conference calls between 6:00 PM and 10:00 PM. Because of this I am more detached from the Pacific Rim, which creates some serious challenges."

Historically, the most effective approach for developing distributed solutions amid complexity has been to employ systems engineering techniques. A *system* is defined as a combination of parts that function as an integrated whole.[17]

Wayne Cascio is a distinguished professor and the Robert H. Reynolds Chair in Global Leadership at the University of Colorado, Denver. His research has identified a number of disadvantages virtual teams face, including the lack of physical interaction, loss of face-to-face synergies, lack of trust, greater concern with predictability and reliability, and lack of social interaction.[16] Each of these disadvantages creates a nuanced challenge for virtual project managers. When aggregated, they create significant complexity.

The information project managers gain from utilizing a project complexity assessment aids in the determination of the skill set and experience level required of the project team. Further, the use of a complexity assessment also helps to guide the implementation of key project processes, such as change management, risk management, and contingency reserve determination; and helps them adapt their management style to the level of complexity of the project.

The project complexity assessment features several parts. The tool includes various dimensions (first part) as defined by a business. Each dimension of complexity is assessed on an anchor scale (second part), and when the complexity scores of each dimension are connected, a line called the complexity profile (third part) is obtained. The complexity profile is a graphical representation of a project's multifaceted complexity.[18] An example of a project complexity assessment is illustrated in Figure 2.6.

Every industry has unique characteristics, every business within an industry is unique, and every project within a business is unique. This means that a firm has to customize the complexity assessment tool for its use. Project managers often start this work by determining the complexity dimensions that are specific to the organization. For example, technical complexity may be related directly to the technical aspect of the product, service, or other capability under development. Structural complexity also has a number of subfactors that involve the organizational elements of a firm. Business complexity involves the business environment in which the firm operates. (See the box titled "Defining Complexity.")

With the dimensions of complexity identified and deemed appropriate for a particular organization and project, the next step in developing a complexity assessment is to define how each

Complexity Dimension	Low Complexity	1	2	3	4	High Complexity
Distributed Sites	Few			X		Many
Communication Technology	Basic			X		Advanced
Culture makeup	Low Variation					High Variation
Project Objectives	Clear		X			Vague
Requirements	Clear		X			Vague
Organization	Hierarchical			X		Matrix
Product Technology	Low Tec		X			Very
Virtual Project Manager Experience	Experienced		X			High Tech Novice
Geography	Local				X	Global
Team Members	Experienced	X				Inexperienced

Figure 2.6: Example Virtual Project Complexity Assessment

dimension of complexity will be measured. This is done by choosing a scale for each dimension.

Alternatives for the scales abound. In the example shown in Figure 2.6, complexity for each dimension is measured on a simple four-level scale (1 is the lowest complexity, 4 is the highest complexity).

Once the complexity dimensions are identified, each dimension is then assessed based on the scale established. For example, in Figure 2.6, cultural make-up of the team is assessed as a Level 2 complexity (medium variation). Once all complexity dimensions are assessed, the obtained scores for each dimension are connected to produce the complexity profile, which helps to visually depict the overall project complexity. The profile in the figure, for instance, indicates that the project is of medium complexity, with all dimensions at Levels 2 and 3, except team members who are experienced (the least complex) and a globally distributed team (the most complex).

Defining Complexity

Complexity arises in many forms. To illustrate complexity, Russ Martinelli and Dragan Milosevic used technical, structural, and business examples in their book, *Project Management ToolBox*.[19] The following is an excerpt from their work that defines each of these areas of complexity.

> Technical Complexity: A feature upgrade to an existing product is low complexity as compared to new product architecture and platform design, which is high complexity. Development of a single module of a system is low complexity as compared to the development of a full system, which is high complexity.

> Structural Complexity: A co-located team is low complexity as compare to a virtual team, which is high complexity. A single site or single geography development is low complexity as compared to a multi-site or multi-geography development, which is high complexity.

> Business Complexity: Selling into traditional or mature markets is low complexity as compared to selling into new and emerging markets, which is high complexity. Flexible time-to-money requirements are low complexity as compared to aggressive time-to-market requirements, which is high complexity.

Every project has a number of variables that determine overall complexity. It is extremely valuable for project managers to understand the complexity of the project, if for no other reason than to mitigate against risk associated with complexity.

Complexity of Virtual Interdependencies

Much of the added complexity of a virtual project is associated with the distribution of work and the web of virtual interdependencies that have to be managed. All projects have interdependencies of work and deliverables. But, for virtual projects, the interdependencies require hand-offs that are distributed geographically and among people who may have little to no personal or professional relationship with one another. This simple fact creates increased complexity of work and therefore increased risk to the project. The best solution to managing such complexity is more frequent collaboration. More specifically, best-practicing virtual project managers know to link collaboration exercises with collaboration practices. They do this by facilitating team development with intentional project planning that necessitates hand-offs more frequently, which in turn necessitates collaboration.

For an example, envision a systems develop project, scoped to take 12 months based on requirements gathering that takes two months, design and wire framing that takes three months, development that takes four months, and quality assurance and testing that takes two months, and a pilot program that takes another month. Also, consider that each function has personnel resources in different locations. As a project manager, you have options and decisions to make regarding the workflow and collaboration of the team. Those managers using best practices won't have business analysts documenting requirements for seven weeks and then hand off those requirements to the design team in week 8. Rather, managers would drive the business analysts to lead more frequent updates, discussions, and collaboration with designers, developers, and the quality assurance and test team. The likely schedule would be once every two weeks. That is the difference between a single hand-off and having multiple points of connection and collaboration.

Similar points of collaboration would be intentionally designed into the project schedule throughout the project's life cycle. Additionally, during these points of intentional collaboration, project managers would integrate team activities so that trust, respect, and rapport are being developed in pace with the project schedule itself. As one virtual project manager explained it: "Rather than focusing solely on the output associated with any single project function, the leader of a distributed project team must focus on creating more touchpoints for the team to come together virtually. Doing this creates more opportunities for providers and recipients of project hand-offs to communicate expectations, which mitigates risk of any hand-off."

With Complexity Comes Risk

There is a relationship between complexity and risk. Virtual projects are inherently more risky than traditional projects. Any of the factors that show up as high complexity in the complexity assessment add risk to a project. The only way to truly understand the risk of any project is to conduct a risk assessment of each project item or variable and determine if mitigation or avoidance actions need to be initiated when a project is initiated and planned.

Project risk is the potential failure to deliver the benefits promised. By understanding and containing project risk, project managers are able to manage in a proactive manner. Without good risk management practices and tools, managers will be forced into crisis management as problem after problem will present itself, forcing teams to react to the problem of the day (or hour). As one well-known author stated, "If you don't actively attack the risks, the risks will actively attack you."[20] Risk management is a preventative practice that allows project managers to identify potential problems *before* they occur and put corrective action in place to avoid or lessen the impact of the risk. Ultimately, this behavior allows the project team to accelerate through the project cycle at a much faster pace. Identifying risk on a virtual project can be especially challenging as explained in the box titled "Virtual Project Risk: Identifying Silent Killers."

Understanding the level of risk associated with a project is crucial to project managers for several reasons. First, by knowing the level of risk associated with a project, project managers will have an understanding of the amount of schedule and budget reserve (risk reserve) needed to protect the project from uncertainty. Second, risk management is a focusing mechanism that provides guidance as to where critical project resources are needed—the highest-risk events require adequate resources to avoid or mitigate them. Finally, good risk management practices enable informed risk-based decision making. Having knowledge of the potential downside or risk of a particular decision, as well as the facts driving the decision, improves the decision process by allowing project managers and teams to weigh potential alternatives, or trade-offs, to optimize the reward-risk ratio.[21]

Virtual Project Risk: Identifying Silent Killers

April Reed and Linda Knight researched the difference in project risk between virtual projects and co-located projects. Their findings were captured in the *Journal of Computer Information Systems*. The following is a summary of their work and findings.[22]

Reed and Knight found 55 risks prevalent with projects today. Of the 55, seven of the risks are specifically noted as being more challenging with virtual teams as compared to co-located teams. These seven specific risks include:

1. Insufficient knowledge transfer

2. Lack of project team cohesion

3. Cultural or language differences

4. Inadequate technical resources

5. Inexperience with company and its processes

6. Loss of key resources

7. Hidden agendas

Some of these risks may seem rather obvious. However, obvious or not, each risk needs attention and mitigation plans. For example, insufficient knowledge transfer is a big problem on virtual teams; knowledge transfer does not occur naturally. Therefore, project managers of virtual teams must intentionally drive knowledge transfer among and between team members. This sharing is associated with the next risk, lack of team cohesion. The more a team can share knowledge and the more it collaborates, the more cohesion it creates. The single most important variable of cohesion is making sure no member of the virtual team feels isolated. It is the responsibility of project managers to drive inclusion and mitigate isolation. Creating this inclusion addresses the potential risk of culture and language differences as well.

Reed and Knight noted:

> Our evidence indicates that the very existence of a virtual project environment makes all these risks more likely to occur and to cause damage. Because none of these seven are traditionally critical risks, they are termed "silent killers." These risks are substantially more likely to occur on virtual projects, yet those who have customarily focused on co-located teams are unlikely to consider them of importance.

The risk assessment and management plan is more than a simple risk register. Importantly, it is a probability tool and response plan. In order for it to work as such, all known and potential risks must be documented, explained by way of project and business implication if realized, ranked in terms of probability, and associated with leading indicators of occurrence to allow as much time as possible to enact a detailed (step-by-step) response plan.

Collectively, that is the risk management plan, and it usually is associated with the planning phase of projects. However, for virtual projects, risk management needs to start in the initiating phase. This is necessary because a good (thorough and accurate) risk management plan can do more than simply illustrate the mitigation of a risk; it can save money and time for project managers and teams. It will certainly be updated and refined in the planning phase and throughout the executing phase, but it needs to start when the project is first initiated.

As can likely be imagined, planning for risk can be much more time consuming for virtual project teams than for co-located teams. When it comes to risk planning, the entire team should be involved. That can occur as a brainstorming session for co-located teams. For virtual teams, project managers often start by emailing each team member, requesting an email returned that contains perceived risks, triggers, impact, and probability of each occurring throughout the project's life cycle. Once the responses are received, virtual project managers aggregate the risks, categorize the risks, and schedules a virtual meeting (or more meetings, as the case may be). In this meeting, it is best if the whole team is involved, but due to geographic distribution, this may not be possible. If it is not possible, project managers will need to conduct multiple meetings.

During the risk management planning meeting, project managers need to be very deliberate with time management as well as the technology used for the meeting. Likely, the technology will be teleconference or video conference. It is best if all participants could have a shared screen so that they can keep pace with the team leader as the team works through each risk. If the project team is distributed as much as illustrated in Figure 2.5, this approach to risk management is the only option project managers have to capture and plan for project risk. In this case, virtual project managers conduct all the same work as co-located project managers, but managing virtual risk is much more difficult and much more time consuming, and require much more planning.

Planning Virtual Communication

The 1967 movie *Cool Hand Luke* made this statement famous: "What we've got here is a failure to communicate!" This statement can very well apply to many virtual projects. A documented communication plan is valuable for any project but is *essential* for virtual projects because of the distributed nature of the team and heavy reliance on technologies to communicate.

Successful projects involve a significant amount of communication among team members. For example, project managers need their management stakeholders to understand project issues. Management needs to understand how a project is progressing and how the team is performing according to its goals and mission. Virtual team members need inputs and information from other team members and reciprocally must provide their inputs on project tasks and issues. Project documentation must be reviewed, discussed, and approved to ensure the project will succeed. To ensure that this level of communication is happening on a virtual project, the virtual team should create a communication plan early in the planning process to solidify their agreement on how information will be communicated among team members and to stakeholders throughout a project.

Communication Items to Be Addressed

A communications plan includes the rules and guidelines by which a virtual project team will manage communications during the life of the project. It is used to help team members think through what kind of communication mechanisms they will need, it helps establish expectations of proactive communication between team members, and it establishes what the team agrees to do. Specific items that need to be addressed in a communication plan include these:

- How status reporting will be performed and who is responsible
- What team meetings will be established and who participates
- What project reviews will be required and who participates
- Who talks to whom and how often

- How documents will be exchanged
- How decisions will be documented and distributed
- What communication technologies will be used

Since project managers and their team members spend much of their time communicating and interacting with one another, a good communication plan that addresses the listed items establishes a framework for effective and efficient communication on the project.

Elements of the Communication Plan

A comprehensive communication plan should include the what, how, and when various forms of team communication and collaboration need to happen. It specifies how the communication between team members and with project stakeholders will be satisfied. Seven key elements of a communications plan are listed next.[22]

1. Electronic technology is the primary communication medium for the virtual team. The team should evaluate and select the most appropriate communication tools available to the organization. Proper training in the applied use of the tools is critically important and must be provided for each participating project location.

2. Principles and guidelines for leading and managing team meetings should be defined. Use good basic meeting management techniques that encourage engagement and frequent team member interaction.

3. Identify all expectations by senior management that lay the foundation with project team members for open communication of information. This will facilitate the flow of project information and issues.

4. Clearly identify the team members responsible for gathering, providing, and distributing specific project information to the project

manager, with the appropriate format and frequency expected.

5. An approved management escalation process and set of procedures is provided for the communication of issues, barriers, and other problems. This process facilitates flow of information to senior managers when higher levels of management attention is needed.

6. Include ground rules for communication of project information such as the use of email, subteam communications, and decisions that are made during information meetings. This ensures that team members have the right access to this informal communication. There should also be appropriate guidelines set for expected maximum response time to emails and voice mails, which can have significant impact on productivity across the project team.

7. If it pertains to a project, the communication plan should also provide the structure for communication to and from outside partners and customers. If the project team involves contractors or outside partners, the plan should also define how they will be involved in team meetings, project monitoring, status reporting, and project reviews.

Using the Communication Plan

Often the communication plan is but one component of the overall project plan. Because of the criticality of effective communication on a virtual project, however, we recommend that the communication plan be a stand-alone document. This will ensure extra focus to increase the probability that the plan contents are put into practice.

On a virtual project, the amount of communication that takes place among team members and with project stakeholders will vary depending on a number of factors, as will the mix of both formal and informal communication exchanges. The two most significant factors that influence the amount of communication are team size and project complexity (where *complexity* is again defined as the number of virtual interdependencies between team members). General guidelines for ensuring the communication plan is effective are listed next.[23]

- *Identify all stakeholders who need to be involved in project communication.* (See Chapter 4.) This will include team members, middle and upper management, external governing bodies, and outside customers, clients, subcontractors, and partners.

- *Select the team communication tools carefully through the development of a technology strategy.* (See Chapter 8.) Communication tools should enable the effective use of the plan, not inhibit it.

- *Decide how often the team needs to communicate and if the communications need to be formal or informal.* Along with this, for formal communications, decide which communications should take place in larger cross-team meetings and which communications need their own forum.

- *Ensure that only the people needed for a particular type of communication are involved.* Often, too many people are involved either by invitation by project managers or by self-invitation. This situation adds risk that miscommunications or misinterpretations will occur. Restrict participation to the critical players needed for an exchange.

The lack of physical presence is one of the major challenges for virtual project teams. Limited physical presence is most commonly thought of as limited face-to-face interaction among team members and with project stakeholders. However, physical presence also includes limited access to critical project information that is needed to ensure that distributed team members view themselves as a cohesive team with a common mission, goals, and plan to go execute. Use of an effective communication plan can assist in overcoming this limitation.

Project planning for a virtual team must be focused on establishing visual representations of what needs to be accomplished, how the team will interact during project execution, and how alignment with to the organization's goals and strategies will be maintained. This planning work, and its associated project plans and deliverables, is vitally important as it creates the foundation and potential for the team to become high performing. Building a high-performing team is the focus of the next chapter.

Assessing Virtual Project Planning

As they navigate virtual projects through the planning phase, project managers must be cognizant of key success factors. Such factors can be discerned from the business case and scope document, among others. The Virtual Project Planning Assessment measures project readiness to move from planning work to executing work. This is, obviously, a very important part of any project since many people perceive the project as actually starting when project execution begins. This assessment can be used as a barometer or benchmark to validate all project plans for adequacy and readiness for project execution.

To perform the assessment, virtual project managers and others on the project team should complete the assessment tool. Once it is complete, they come together to discuss responses to each item and to mitigate any "no" response that may prevent the team from achieving project outcomes.

As can be seen, when each of the items in this assessment is answered "yes," the organization (specifically the project manager managing the project and the stakeholders impacted by it) are more apt to be "ready" than not. Whenever there is a "no" response, it warrants further discussion to determine why and is a prompt to further detail project plans or steps to put into place to move the "no" to a "yes" response.

Virtual Project Planning Assessment

Project Name: _____

Date of Assessment: _____

Assessment Completed by: _____

Assessment Item	Yes or No	Notes for All No Responses
Project Setup		
The project mission and vision are documented.		
Project goals and objectives are defined and documented.		
The project statement of work and scope are clear and documented.		
The project success factors and key performance indicators are defined and documented.		
The project requirements are clear and documented.		
The project business case has been approved.		
The project charter has been documented.		
A project complexity assessment has been performed.		
Project Team		
The project architecture has been documented.		
All distributed team sites have been identified.		
All external partners have been identified and are participating in the project planning as required.		
The project team charter has been documented.		
The project mission, vision, and goals have been communicated to and reviewed by all project team members.		
All tasks have been adequately distributed to the various team sites.		
The experience level of the team is consistent with the complexity level of the project.		
A cultural complexity assessment has been completed.		
There is a high level of change acceptance (low level of change resistance) in the organization.		
There is a high level of leader credibility (of sponsor and project manager) in the organization.		
The organizational decision making process is participative and collective.		

Assessment Item	Yes or No	Notes for All No Responses
Project Plan		
The project deliverables and key milestones have been mapped.		
Each distributed project team member has completed their task plan.		
Each distributed team member has a baseline schedule.		
Each distributed team member has an estimated budget.		
Each distributed team has a resource plan and commitment for resources from functional managers.		
The integrated project plan is complete.		
The integrated project plan supports attainment of the project goals.		
The integrated project plan has been approved and organizational commitment has been received.		
Technology		
A technology strategy has been completed for the project.		
Virtual communication and collaboration tools have been selected.		
Local infrastructures at the project teams' sites will support data performance requirements for the selected technologies.		
Virtual communication and collaboration tools have been implemented.		
All team members have access to the communication and collaboration technology platforms.		
All team members have been trained on use of the technology.		
Execution Strategy		
The project and expected outcomes have been considered and discussed as a priority for the organization.		
The virtual project is planned to integrate cross-team collaboration and allow for and celebrate short-term wins.		
There is a mature method for managing the project and leading a geographically distributed team.		
The communication plan is detailed and appropriate relative to the project complexity.		
The virtual project team understands and has approved the communication plan.		
The communication plan clearly details an escalation process for conflict management and decision making.		

Findings, Key Thoughts, and Recommendations

Notes

1. John Maxwell, *The 360 Degree Leader: Developing Your Influence from Anywhere in the Organization* (Nashville, TN: Thomas Nelson, 2011).

2. Project Management Institute, *A Guide to the Project Management Body of Knowledge*, 5th ed. (Newtown Square, PA: Author, 2013).

3. www.dictionary.com.

4. Russ J. Martinelli, Jim Waddell, and Tim Rahschulte, *Program Management for Improved Business Results*, 2nd ed. (Hoboken, NJ: John Wiley & Sons, 2014).

5. Martinelli, Waddell, and Rahschulte. *Program Management for Improved Business Results,* 2nd ed.

6. Peter M. Senge, *The Fifth Discipline: The Art and Practice of the Learning Organization* (New York, NY: Doubleday/Currency, 1990).

7. Russ J. Martinelli and Jim Waddell, "Demystifying Program Management: Linking Business Strategy to Product Development," *Product Development Management Association* (January 2004): 20–23.

8. Geoffrey A. Moore, *Crossing the Chasm: Marketing and Selling Technology Products to Mainstream Customers* (New York, NY: HarperBusiness, 1991).

9. Project Management Institute. *A Guide to the Project Management Body of Knowledge.*

10. Russ J. Martinelli and Dragan Z. Milosevic, *The Project Management ToolBox*, 2nd ed. (Hoboken, NJ: John Wiley & Sons, 2016).

11. Project Management Institute. *A Guide to the Project Management Body of Knowledge.*

12. Cristina B. Gibson and Susan G. Cohen, eds., *Virtual Teams that Work: Creating Conditions for Effective Virtual Teams* (San Francisco, CA: Jossey-Bass, 2003).

13. Ken Ferrazzi "Getting Virtual Teams Right," *Harvard Business Review* (December 2014). https://hbr.org/2014/12/getting-virtual-teams-right.

14. *New Oxford American Dictionary*, 3rd ed. (New York, NY: Oxford University Press, 2010).

15. Martinelli, Waddell, and Rahschulte. *Program Management for Improved Business Results.*

16. http://www.worldtimezone.com.

17. R. Stevens, *Systems Engineering: Coping with Complexity* (London, UK: Pearson Education, 1998).

18. W. F. Cascio, "Managing a Virtual Workplace," *Academy of Management Executives* 14, no. 3(2000): 81–90.

19. Martinelli and Milosevic. *The Project Management ToolBox.*

20. T. Gilb, *Competitive Engineering: A Handbook for Systems Engineering, Requirements Engineering and Software Engineering* (Oxford, UK: Butterworth-Heinmann, 2005).

21. Martinelli and Milosevic, *Project Management ToolBox.*

22. A. H. Reed and L. V. Knight, "Project Risk Differences Between Virtual and Co-Located Teams," *Journal of Computer Information Systems* 51, no. 1 (2009): 19–30.

23. M. R. Lee, *Leading Virtual Project Teams: Adapting Leadership Theories and Communications Techniques to 21st Century Organizations* (Boca Raton, FL: CRC Press, 2014).

BUILDING A HIGH-PERFORMANCE VIRTUAL TEAM

It is well understood that the mission of any project is to create and deliver a solution that advances the business goals of the sponsoring company. It is also understood that there are two primary parts to any project: *managing* the project process and *leading* the project team. These fundamentals of project work exist in both traditional and virtual environments. However, how the project work is performed by the project team differs between the two environments.

As described in Chapter 2, management of the project process through a project management methodology with its associated practices, processes, and procedures does not change significantly between traditional and virtual projects. All projects must be initiated and planned to a certain scope of work and must be managed from cost, schedule, scope, risk, and change perspectives. In addition, work outcomes must be integrated to create a solution and then effectively brought to closure. Leading the project team, however, can vary vastly between traditional and virtual projects.

In essence, project team members perform their work in two ways: individual work and teamwork.[1] People perform their individual work in identical fashion whether they are part of a traditional project or a virtual project. How they perform their work within the context of the team can differ significantly between the two types of projects, however. Effective teamwork is dependent on effective cross-member collaboration. Effective collaboration is in turn based on the establishment of professional relationships, trust between team members, clear roles and responsibilities, and appropriate communication forums to name a few of the important factors. For project managers of virtual teams, establishing effective teamwork has little to do with the science of project management and the *hard* project management skills they have developed and honed. Helping the virtual team collaborate in a new virtual paradigm relies heavily on the *soft* project manager skills, as they are often called, which include the art of leadership and the people skills that aren't learned and certified so easily. As managers know, the science of project management will fail quickly if project team interactions become dysfunctional. Project managers may realize early in the project cycle that the people skills that worked effectively on traditional project teams do not work as well for virtual teams. In fact, the separation of distance, time, and sometimes language and culture that defines a virtual team often requires project managers to focus more on team leadership and less on the fundamental science of project management. At the very least, project managers must have a clear understanding of the two distinct roles they must play—the role of project manager and the role of project team leader.

This distinction between the two roles of virtual project managers is not lost on Jeremy Bouchard. "While managing my first virtual project, it occurred to me that I overemphasized my knowledge and understanding of project management and that I woefully underestimated the importance of being an effective leader of the project team," he explains. "For most of my career as a project manager, I've been very focused on project management methodologies and processes. You just sort of get on that track once you immerse yourself in project management. I let myself believe that my project management credentials were enough to carry me from project to project. All was good until I stepped into the world of virtual projects."

Fortunately for Bouchard, he works for a person who understands that project management credentials are like table stakes that get you into a game of poker. They are necessary, but insufficient. His manager and mentor, Brent Norville, helped Bouchard learn that the balance of effort on a virtual project shifts from project management fundamentals to team leadership fundamentals. "It takes a while for a project manager with experience managing traditional projects to come to the realization that the project team has to be functional and perform as a unit in order for project management processes to be most effective," Bouchard explains.

As Bouchard reflects, "Fortunately, my first virtual project wasn't very large or complex, so I was able to navigate through it with some great coaching from Brent. I now know that my team leadership skills and experience were weak. I still have a lot to learn, but now I understand the importance of the leadership role of the project manager. On my current project, which is also virtual, I have spent much more time building my team and focusing on being a strong leader. In the process, I've had to learn to delegate."

It bears repeating that success in managing virtual projects requires a high level of skills and competence in both project management and project team leadership. We described how the fundamentals of project management must be nuanced to translate into the world of virtual project management in Chapter 2. In this chapter, we turn our attention to the important nuances pertaining to the leadership of a virtual project team. We begin by looking at the most common types of virtual project teams.

Virtual Project Team Types

When people discuss the essentials of leading a project team, especially a virtual project team, much of the attention is centered on effective communication. Although communication is an essential part of all team dynamics, attention to people's feelings, priorities, and perceptions is also important, especially when we are trying to build a high-performance team.[2] The managers of traditional projects have a distinct advantage when it comes to project leadership and the people side of a project because they interface directly with their team members and can rely on visual observations to determine personal and performance characteristics. This is not the case for virtual project managers.

Virtual project managers quickly realize that the people skills that work well for traditional projects may not work as well on virtual projects. Traditional approaches must therefore be modified. The level of modification is dependent on the type of virtual team that is being managed. For this reason, we briefly describe five common virtual project team types:

1. Mostly co-located, one central location
2. Mostly co-located, multiple national locations
3. Mostly co-located, multiple global locations
4. Mostly virtual, nationally distributed
5. Mostly virtual, globally distributed

Mostly Co-located, One Central Location Model

Becoming a virtual organization is often a journey, and many organizations begin with a virtual project team model where most of the team is concentrated in one or two locations with several members working from remote locations. (See Figure 3.1.)

This model is usually the result of hiring a few employees with very specialized skills who are located outside of the corporate geography and have reached agreement with the company to remain in their locations.

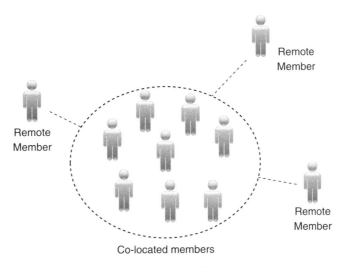

Figure 3.1: Mostly Co-located Virtual Project with One Central Location

Because the large majority of team activity is concentrated at the centralized location, the remote team members face a constant challenge of staying tightly connected to the project as the large majority of team interactions are based on face-to-face exchanges between the co-located team members.

Building a high-performance team in this model is relatively similar to building a fully co-located team. The challenge, of course, is to ensure inclusion of the remote team members. Special attention and time is required to make sure that remote members fully identify as being part of the project team and interact in a seamless manner with the other team members, even though they have to do so using technology.

Mostly Co-located, Multiple National Locations Model

In the mostly co-located with multiple national locations virtual project team model, there are several geographic concentrations of team members, but all locations are in a single country. This model often emerges as a result of company acquisitions, business scale-up, or reorganization endeavors where multiple business centers are formed at the corporate level. Over time, work begins to integrate between business centers, and mostly co-located, multiple location project teams emerge. (See Figure 3.2.) It is also not uncommon for a few remote members with specialized skills to be part of the virtual project as well.

The difficulty in building high-performance team increases with this virtual project team type due to the separation of organizational business units or divisions underlying this model. As stated earlier, this model often emerges as a result of a company acquisition or reorganization. Consequently, each location normally has its own management team, structure, and local business culture (*culture* being defined as "the way we do things" in this context). The difficulty, therefore, comes in building a common project team identity with team members who already possess a strong, but differing, organizational identity. In addition, project managers have to intentionally and purposefully work to build strong working relationships with key management stakeholders at each business location. A high-performance team cannot be built without the support of the organizational managers. These individuals will either be enablers or distractions to building the team.

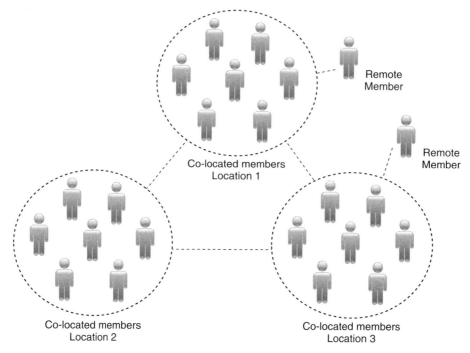

Figure 3.2: Mostly Co-located Nationally Virtual Project with Multiple Locations

Mostly Co-located, Multiple Global Locations Model

Most companies today realize the benefits of conducting business on a global scale (or at least realize the competitive disadvantage of not doing so), so many business partnerships and business expansions occur in various parts of the world. This creates a global element to virtual project teams. In mostly co-located, multiple global locations virtual project teams, there are several locations of project team concentration in various parts of the world, as demonstrated in Figure 3.3.

The ability to build a high-performance project team in this virtual team model is hampered by differing cultural factors and large separation in time and distance between team members. Creating a common team identity becomes even more important in this model, but requires significantly more work on the part of project managers. Likewise, ensuring that effective team interaction is occurring across organizational, cultural, and national

boundaries can be nearly a full-time job in itself; yet it is a job that is foundational to project success.

Mostly Virtual, Nationally Distributed Model

In the mostly virtual project team models, team members are geographically distributed in relatively equal fashion, as illustrated in Figure 3.4.

Some locations will include a larger concentration of team members than others, but it is not significant enough to tip the balance of team interaction to a mostly co-located model. What is prevalent instead is the lack of team member concentration found in the mostly co-located virtual project team models. Because project members are not concentrated in a few locations, building a high-performance project team is in many ways easier because members are in parity without being encumbered by organization and management hierarchy issues. Team members often seek team identity as a substitute for weak organizational

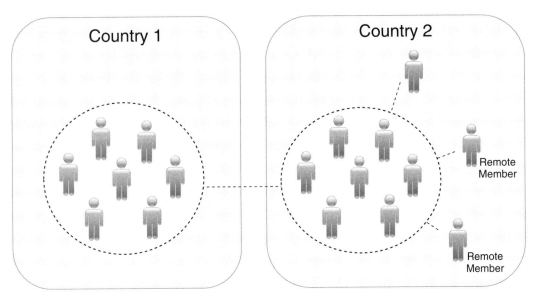

Figure 3.3: Mostly Co-located Virtual Project with Multiple Global Locations

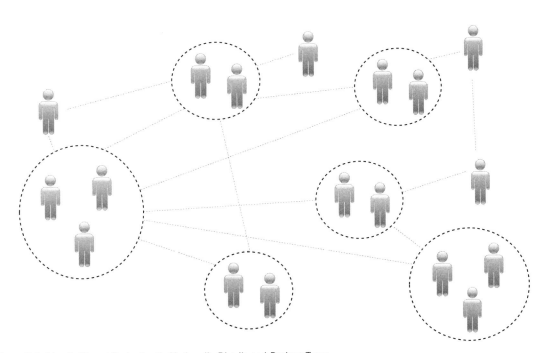

Figure 3.4: Mostly Virtual Project with Nationally Distributed Project Team

identity and tend to look for coworkers who share a common purpose for their efforts and contributions to the organization.

Geographic separation limits face-to-face interactions; therefore, communication, collaboration, and relationship building has to occur via technology. And importantly, cross-team interactions normally have to be facilitated at least initially by virtual project managers. Again, ensuring that the interactions occur requires more time and attention.

Mostly Virtual, Globally Distributed Model

Adding the global element to the mostly virtual project team model brings with it the complexities associated with multiple national cultures, multiple first languages, and a greater degree of separation in time and distance between team members. This model is very much a hybrid between the mostly co-located, globally distributed and the

virtual, nationally distributed project team models that combines both the benefits and challenges of the two models. (See Figure 3.5.)

Because team members are not strongly connected to a centralized location, they are often eager to establish a connection and identity to a project team. In many cases, the project team identity becomes their company identity.

However, significant effort on the part of project managers is required to initiate team interactions and to facilitate cultural nuances that will occur within the team. As we cover in detail in Chapter 8, selection of the right technologies that enable effective team communication, collaborative interaction, and relationship building is critically important and largely the responsibility of project managers.

Much has been written about achieving high-performance teams. No doubt, achieving high performance is a combination of both management and leadership factors regardless of whether the

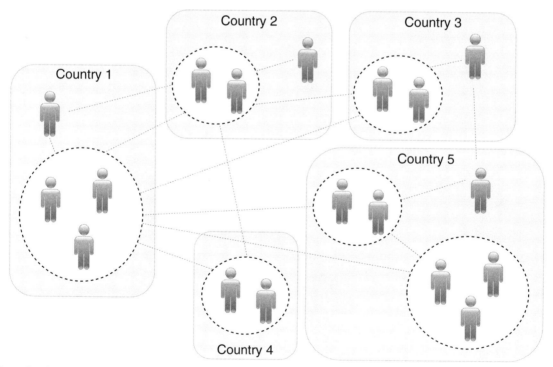

Figure 3.5: Mostly Virtual Project with Globally Distributed Project Team

project team is traditionally structured or has some type of virtual structure. Those enterprises that can achieve higher-performing virtual project teams will possess a significant competitive advantage. But what does it take to build a high-performing team? Answering that question must start with a discussion on the foundational elements of teamwork.

Differentiating High-Performance

Let us begin by distinguishing between a work group and a team. Both a work group and a team can be defined as a group of people working together to accomplish an objective or goal. The difference between the two comes in understanding the interaction between people. A work group can achieve its objectives by its members working separately and independently, such as the members of a project management office. By contrast, a project team provides a collective work product and a performance gain or improvement that is not achievable by group members working on their own. If it is a project team, then it is represented by a set of team members working under the direction of a project manager and performing project work to achieve the project objectives.[3]

How effective a project team is depends on how well members work together in creating the work product and achieving the project objectives. This is commonly known as teamwork. A good working definition of *teamwork* is "the process of working collaboratively with a group of people in order to achieve a common goal."[4]

Good teamwork is rooted by a set of shared values that enables important behaviors including listening and constructively responding to the points of view of other team members, providing support to each other on the team, and being positive and supportive to the accomplishments of others on the team.[5] Teamwork is further encouraged and achieved during the life of the project

through good communication and collaboration practices, mutual respect, appropriate processes, and clarity in decision making procedures.

Not all project teams are created equal. All projects consist of a team of people who perform the work intended. *Some* project teams seem to consistently perform at a higher level than others. A high-performance team is one that consistently exceeds the expectations of customers, sponsors, and senior management. Normally, what comes to mind when most of us think of high-performance teams is a sense of accomplishment that goes beyond expectations. Jon Katzenbach, a leading author and practitioner in organizational strategies, accurately describes a high-performance team as "a group that meets all the conditions of real teams, and has members who are also deeply committed to each other's personal growth and success. That commitment usually transcends the team and the team has internalized the philosophy that if one of us fails, we all fail."[6] It is not easy to assemble and create a high-performance team however, and most organizations struggle to build teams that can consistently exceed expectations. Teams of this caliber are rare.

High-performing teams develop a keen sense of shared accountability and responsibility for the overall success of the project. This sense of joint responsibility is interpreted as not only fulfilling their own individual and personal commitment to the team, but also making sure that all team members are successful. Shared responsibility works best if everyone on the team is clear about their own responsibilities, as well as joint responsibilities, and fully understands the success factors and objectives of the project. Of course, high-performance teams do not just form by themselves. All high-performance teams share a common element: a project manager who also performs as an exceptional team leader. High-performance teams have to be built and sustained, and those that achieve high-performance status are led by people who have a good balance between project management and project leadership capabilities.

In short, high-performance teams are intentionally designed and well led.

What prevents all project teams from becoming high-performance teams? Some believe that if you add your top talent and best performers to a team, it can't help but be successful—the "dream team," if you will. Most of us have seen that this rarely works on its own. Usually some underlying fundamental principles of team building have been ignored or left unaddressed when forming the team. In other cases, organizational issues or barriers may exist that hamper or do not enable the team's ability to perform.

Many authors have tackled the subject of characterizing high-performance teams. Probably two of the foremost experts are Jon Katzenbach and Douglas Smith. In their book *Wisdom of Teams*, they cite five key criteria possessed by teams performing at a higher level than average or even good teams. These include:

1. A deeper sense of purpose
2. Relatively more ambitious goals
3. Better work approaches
4. Mutual accountability
5. Complementary skill set (and, at times, interchangeable skills)

They point out further that high-performance teams have a unique quality in that the team members have a basic need and an ambition to go after bigger challenges. Consequently, they bring with them a work ethic that creates a deeper commitment to the collective mission.[7]

As a result of our own experience leading project teams, we feel there are a number of other critical characteristics exhibited by high-performing teams that should be pointed out:

■ The presence of a shared vision and objectives

■ Participative leadership

■ Well-defined roles, responsibilities, and expectations

■ Action-oriented team members

■ Trust between team members

■ Managed conflict

High-performance project teams can be a marvel to observe. It is evident, even to casual observers, that members of high-performance teams possess a passion and energy for exceptional work outcomes. It is this bias for excellence that binds employees together across both hierarchy and geography and guides them to make the right decisions and advance the business without explicit direction to do so.[8]

Building a High-Performance Virtual Project Team

Despite the ever-present challenges to building a virtual project team that performs at a high level on a consistent basis, there is good news for virtual project managers. Many of the foundational elements of building a high-performance team are the same for both traditional and virtual teams. The major difference is in how these elements have to be applied when the team is geographically distributed.

We admire the quote by the British statistician George Box, who stated, "All models are wrong, some are useful." One such model that we find tremendously helpful in building a high-performance virtual project team is what is commonly referred to as the Tuckman development model. Defined by Bruce Tuckman in 1965, the model describes four stages of development that a team normally progresses through: forming, storming, norming, and performing. Although decades old, this model of team development still serves as an excellent base for understanding what must transpire in the establishment of a virtual project team (see box titled "Challenges to Developing a High-Performing Virtual Team"). In particular, Tuckman's forming and storming phases offer considerable understanding as to how teams evolve in the early stages of a project and establish the foundation for high-performance later in the project.

Challenges to Developing a High-Performing Virtual Team

Psychologist Bruce Tuckman mentioned the need for teams to navigate through forming, storming, norming, and performing in his 1965 article, "Developmental Sequence in Small Groups." This is the only path to high performance. Researchers Stacie Furst, Martha Reeves, Benson Rosen, and Richard Blackburn identified best practices for leading virtual team development through the Tuckman stages in their article "Managing the Life Cycle of Virtual Teams." Some of their findings regarding the challenges unique to virtual teams when it comes to achieving high performance are described next.[9]

Forming. There are fewer opportunities for conversations, especially informal work and non-work-related conversations. Because of this, there is increased likelihood of stereotyping fellow team members and making decisions from assumptions rather than facts. Further, with the absences of accurate information, trust is slower to development among team members. Therefore, it becomes absolutely critical for the virtual project team leader, during the formative stage of the project, especially with limited to no face-to-face time and involvement allowed by the organization, to ensure that a sufficient number of virtual team meetings (video/audio conferences, etc.) are held. These meetings should be orchestrated by the virtual team leader with well-planned agendas and time allowed for team members to get to know one another and discuss project vision, objectives, team charter, team norms, the project team communication plan as well as other pertinent and related topics.

Storming. With fewer communication channels, conflicts can become exacerbated and poor performance can evolve because of passive-aggressive behaviors among team members. Further, team members can more easily withdraw from the collaboration tasks and activities, especially during this phase of team development. These adverse behaviors can be reduced significantly by consistently adhering to and applying the team norms, which should be discussed, established, and agreed on by all team members during the project forming stage. Included should be procedures and processes for running team meetings, communication protocols for each communication, collaboration tools the team will utilize, and established norms for team and one-on-one interactions specifying acceptable behaviors between team members.

Norming. The speed of communication and lack of speed in response to communication can cause frustration and limit commitment among team members. The anticipated and estimated time loss and delays created by using electronic means for all or most communication and collaboration across the multiple distributed site locations need to be properly factored in to the virtual project schedule. Potential adverse impact to project team member responsibilities and deliverables, many times, can be due to lack of response of other team members. This is a management and performance problem that needs to be consistently monitored and appropriately addressed by the virtual project manager when it occurs. Project managers who have experience leading virtual teams know that these types of problems can be anticipated, due to the nature of distributed teams, and procedures for handling them must be agreed upon by all team members. This should be accomplished during the formative stage of the project.

Performing. Often team members are assigned to multiple teams, which can create competing pressures for virtual project managers and local managers. Experienced and knowledgeable organizational and project managers understand the risks, barriers, and time commitments faced by project personnel assigned to more than one project. The local site managers and virtual project managers are the best ones to be held accountable to oversee and ensure that critical project resources are not overcommitted to project work that exceeds people's ability to perform and accomplish their assigned project workload. When projects start significantly missing intended targets and performance objectives, this is an indicator that project resources are not allocated effectively.

The forming stage is especially critical as this is the time when team members get acquainted with one another and gain knowledge and understanding of the expectations for their participation on the team. Team members are also eager to learn about the project requirements and deliverables.

In Tuckman's storming stage of team development, team members begin to understand their respective roles, responsibilities, and dependencies on each other and how they might, as individuals, need to bend and change their ideas, attitudes, and approaches to adhere to the group.[10] Conflicts may begin to occur, which gives rise to the opportunity to begin directly working through the conflicts, a fundamental element of high-performing teams. We suggest that all virtual project managers either familiarize or refamiliarize themselves with the early stages of the Tuckman model. It can serve as a good tool for establishing a firm understanding of the team dynamics that are present during the formation of every virtual project team.

Effective Team Formation

Team formation for virtual project teams is particularly challenging. Project managers of traditional projects have a distinct advantage over their virtual project counterparts in regard to forming their project team. For traditional projects, team leaders often are aware of the capabilities and personalities of many, if not all, individuals who are candidate team members. Additionally, these leaders have had an opportunity to build relationships with the functional managers who decide who will be assigned to what project and therefore can influence those decisions. Virtual project managers rarely possess these advantages and must try to form high-performance teams without knowing the capabilities and qualities of the people assigned to them.

One of the foundational elements of a high-performance team is mutual trust, both among team members and between team members and the project leader. The limits of the technologies used in the communication and collaboration needed between geographically separated team members normally constrains building the much-needed trust and shared understanding of all aspects of the project. Research has shown that relationships, trust, and team identity can be effectively created and formed on virtual project teams, but at a slower pace than for co-located teams. There is no doubt that it is considerably easier and faster for co-located teams to build relationships using face-to-face meetings and interactions. Most virtual teams have limited to no access to this major team-building advantage.[11]

Teams also rely on understanding and internalizing the various sources of project information as part of the team formation process. A traditional team has a high level of social presence that facilitates the sharing of information, especially in the forming of the team. It is much more difficult in the virtual team environment to share and gain access to virtual project team data and information. Considerably more effort and effective electronic systems are required to have project documentation and other information readily available to all team members. Sufficient discussion must occur to reach a common acceptance and internalization of project information as a team. The next box, titled "Eva's Virtual Project Team Formation Template," is an example of how one virtual project manager approached the formation process for her project.

Eva's Virtual Project Team Formation Template

Eva Garcia is the virtual project manager for a new information technology effort for her globally distributed firm. She is leading her newly appointed team from her home office in Vancouver, BC, Canada. Earlier, when the project was approved by senior management, she negotiated with her general manager and functional managers for the best team personnel available to be assigned to her project.

Her virtual team members reside at the following sites of her company: Vancouver, BC; Dublin, Ireland; Madrid, Spain; and Taipei, Taiwan.

This is Eva's second globally distributed virtual project. She is preparing to lead the first of a few virtual meetings targeted at the team formation process via a scheduled video conference with all team members present. Eva, an experienced project manager, has learned from her past projects, and especially her first virtual project two years ago, that the process of team formation is critical to having the highest probability of achieving the intended project goals. Eva strongly believes that even if you have the best talent and best-in-class tools for your project, you will not achieve success on the project if the very important steps of team formation are not done properly.

From her experience, she created a systematized virtual project team formation template of activities (represented in Figure 3.6), which assisted her in the past and she plans to follow it again on her new virtual project. Eva knows that these steps to team formation are critical in helping her to build the trust, chemistry, and team relationships required. Team members need to be clear on what the purpose of the project is, what is expected of them, what their role is, and how the team will operate and govern itself. Of course, she also realizes from her past traditional projects that these team formation tasks will be much more difficult to accomplish in the virtual environment due to the time, distance, language, and cultural barriers, which result in more meeting time to work through the formation tasks.

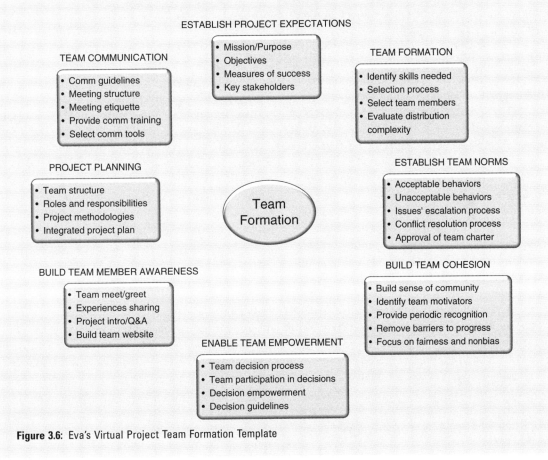

Figure 3.6: Eva's Virtual Project Team Formation Template

Eva is also very aware that she must be the key driver on the team for building and instilling trust. She believes that the foundation of building trust with team members must be based on her competence as a leader, not just her assigned role as project manager. She knows that her effectiveness will hinge on her integrity and by being consistent in exhibiting her commitment and concern regarding the needs and well-being of her fellow team members.

Further, she knows and believes that if virtual project team formation is performed effectively, it will lead to a productive team that works well together.

Creating a Common Purpose

Bringing a group of people together and assembling them as a team does not make them think and behave as a team. Members of a team must view their work in terms of *we* instead of *me*; in other words, they must think: *We* must work together toward a common purpose that is defined by a set of common and agreed-on business and project goals. Project managers must establish the common purpose and inspire the team to work collaboratively to achieve the goals that define that purpose. The ability of project managers to create a common vision and the team's willingness to adopt that vision is what defines a group of people as a team. For a virtual project, the need for a well-defined common purpose is amplified because it is the foundation on which building team cohesion is based.

When a virtual project team is formed, team members usually have very little in common. At times, as in the case of a merger or acquisition of companies, team members may not have worked together previously. Additionally, in a merger or acquisition, team members have very idiosyncratic ways of thinking and behaving within the cultural and functional norms of their own company. At times they place their own personal goals ahead of team goals. Without a clearly articulated common purpose for the project and project team, a large degree of ambiguity and lack of focus can exist within virtual teams. A clearly articulated common purpose is a virtual project manager's most valuable tool for driving ambiguity out of the environment and getting team members to think of team goals ahead of personal goals.

A clearly defined common purpose should answer four key questions:

1. What is the purpose of this team—the mission?

2. What does the end state look like—the vision?

3. What do we need to accomplish to get to the end state—the objectives?

4. How will success be measured—the success factors?

The job of virtual project managers is to create answers to these four foundational questions with input from team members and stakeholders. Answers should be simple, direct, and free from jargon to overcome competing cultures and mind-sets and to address both corporate needs and local conditions. Creating the answers to these four questions is the first step toward getting team members to think and behave collectively instead of individually.

In Chapter 2, we introduced the whole solution as a means for project managers to establish both a common mission and a vision for their project teams. The whole solution is defined simply as "the integrated solution that fulfills the customers' expectations."[12] *Integrated* is the key word in this definition. This word tells us that customers' expectations cannot be fulfilled by any one specialist or set of specialists on the team. Rather, success comes when meeting customer expectations is a shared responsibility between project team members, with their work tightly interwoven and driven toward an integrated customer solution.

By creating the whole-solution diagram, a visual representation of the integrated solution that will meet the customers' expectations emerges. An effective illustration of the whole solution occurs when each member of the project team can see how their work contributes to the creation and delivery of the integrated solution. The whole-solution diagram also allows project team members to see what work is done where, geographically. In the example depicted in Figure 3.7, product software and circuitry is provided by the team in Palo Alto, California; the power supply team is in Dallas, Texas; enclosure work is performed by the New York team; platform and infrastructure is managed by the Ireland team; radio and other telephony services is provided by the Sao Paulo team; manufacturing is conducted in Shanghai; and quality control and customer support is the responsibility of the Sydney team.

A whole-solution illustration becomes central to establishing a common vision and serves as a guidepost for project team members to understand what work the team collectively needs to do on the project. When a team is distributed and works virtually, it has a more difficult time seeing itself as a whole that is working collaboratively toward common goals. Each team member needs to be able to see how their actions and those of others contribute to goals—how individuals impact the collective.

The whole solution, however, is merely the means to achieve the business results that are driving the need for the project. Clear definition of the business results form the answer to the third question in creating a common purpose—what are the project objectives? In the example above, the project objectives may be to increase market share, increase profit margin, lower product cost, and accelerate time-to-market introduction. By defining and documenting the project objectives, members of the virtual project team begin to focus on the things that can be achieved only by the collective success of the team.

It is not sufficient to define just the project objectives when establishing a common purpose for a project. The final step comes in defining the project success measures. Specifically, the success measures must define how the project objectives will be measured. For the example used previously, project success may be measured by an increase of 5% market share, 2% increase in profit margin, $50 decrease in product cost, and market introduction in the fourth quarter of the year. As with the objectives themselves, it is clear that the success measures can be achieved only by the collective success of the project team.

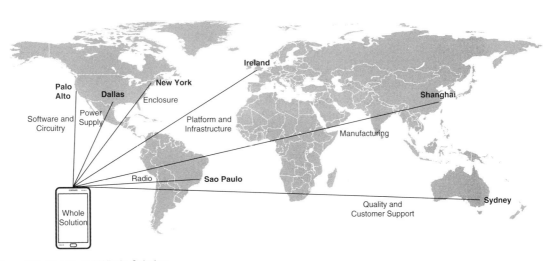

Figure 3.7: Distributed Whole Solution

Creating Team Chemistry

A project team may consist of the top talent within an organization, but they will not reach a high level of performance without a certain bonding of spirit and purpose. It is this bonding, or team chemistry, that motivates team members to work together collaboratively for the common success of the team.

Of course, when a group of people form a team, their personalities do not gel immediately. Acculturation of personalities, ideas, shared values, and goal alignment takes time as well as intentional effort (sometimes considerable effort) on the part of project managers.

People from diverse backgrounds and experiences will bring different behaviors, routines, values, and ideas about the work of the team. Team leaders must embrace this diversity of people on the team as individual members who make up a collective work unit, and they must act as coaches and role models for the rest of the team to help them embrace the value of diversity. Chemistry will evolve from how well each team member bonds to others on the team and how much each is willing to contribute to make the team a success. No doubt, project teams with excellent chemistry between the team members are much more productive and achieve higher levels of performance.

Establishing team chemistry on a virtual project is complicated because of cultural diversity, lack of social presence, and increase in communication challenges due to language and time zone barriers brought about by the team's geographic distribution. As a result, when team members represent diverse backgrounds, converging national, organizational, and functional cultures into a team culture where values, attitudes, and meanings come together takes more time. On a widely distributed team, it is also more difficult to get to know people on a personal basis and to form close relationships due to the limited amount of ongoing face-to-face interaction. However, both the blending of culture and development of personal relationships are critical for the team to behave cohesively and consistently *as a team* instead of as individuals.[13]

Successful virtual project managers do a number of things to accelerate the establishment of team chemistry. These include establishing clear roles and responsibilities, fostering social presence, using information-rich communication technologies, and celebrating team successes.

Establish Clear Roles and Responsibilities

A project team cannot achieve a state of high performance without a clear understanding of who is responsible for what. For traditional projects, this clarity often can be achieved through informal and formal conversations. That is not the case for virtual project teams. Team member roles and responsibilities have to be explicitly documented and communicated on the virtual project; a simple responsibility matrix, as shown in Figure 3.8, is often all that is needed.[14]

Ultimately, virtual project managers are responsible for the various outcomes and outputs associated with a project as well as the successful achievement of the success factors. However, along the way, responsibility for the satisfactory completion of work is a shared responsibility between the

	System Designer David P.	Marianne V.	Aaron M.	Bridget B.	Victor P.	Gina C.	Siva G.
HW Solution		A	C				
HW Definition		R	C				
HW Requirements	A	R	C				
SW Platform		R			A	C	
SW Architecture		R			A	C	
SW Requirements					A	C	R
Applications				C	R	C	C
System Security	A		C	R			
Infrastructure Definition	A	R	C				C

A: Approver   R: Reviewer   C: Creator

Figure 3.8: Example Project Responsibility Matrix

project team members. (See the box titled "Virtual Roles and Responsibilities.") A best practice among virtual project managers is to use a responsibility matrix to explicitly demonstrate how responsibility will be delegated and shared on the project to remove implied assumptions on the part of the project team members and stakeholders about *who* is to do *what*.[15]

Virtual Roles and Responsibilities

Sylvie Huyskens, a senior information technology project manager, shared with us that on one of her virtual projects, she led a European roll-out of human resource software with technical support from the United States. "Of course, I was apprehensive about the difference in languages and culture, which might cause issues of the 'lost in translation' kind. However, our first hiccup turned out to be roles and responsibilities. We had people assigned to the project with the same role, which would impact decision making and collaboration. So, we talked, clarified assumptions, and figured it out. We went through all the roles, just to be sure we had no overlap. The project turned out to be highly successful. I learned that taking a step back to ensure everyone has the same understanding about their role on a project guarantees better results."

Foster Social Presence

To prevent some members of the virtual team from becoming invisible, make sure that all team members know one another and continue to foster connections. The best scenario for developing and establishing these foundational elements, given the time and distance barriers associated with distributed virtual teams, is through one or more early face-to-face team meetings. Unfortunately, many virtual teams have little to no face-to-face team involvement, which can limit and slow effective virtual team development. It is important for senior leaders to realize that investing company financial resources to bring teams up to speed more rapidly and to increase the potential for achieving higher levels of virtual team performance is well worth the investment.

A recent study by Deloitte pointed out that "virtual teams should not discount the importance of face-to-face interactions with the client and the onsite team. Teams may consider rotating offsite resources onsite to enable them to have client-facing interactions and get to know the rest of their team in person." Deloitte further pointed out that after the rotation, these resources typically are more effective due to the deeper relationships that have been formed.[16] Many organizations have found it beneficial to invest initially in face-to-face team meetings during early team formation and also provide the opportunity to spend some time on building relationships.[17]

However, face-to-face team meetings are not always financially possible. The fallback position is to accomplish these early formative interactions through audio or video conference sessions. These sessions, of course, must be followed up with periodic visits and meetings on a consistent basis between project managers and team members. It is important for project managers to set the tone and perspective by getting face-to-face with team members. Doing this allows team members to visualize project managers and create a visual frame of reference that will carry forward and be projected on future audio and written communications. Nothing can fully compensate for the inability to see a person's body language while communicating with them, but having a visual frame of reference helps tremendously in furthering social presence with a team.

With face-to-face interactions being rare on virtual projects, the team must be able to use communication and collaboration technologies that enable them to continue building relationships and rapport. Video conferencing, either roombased or desktop based, is a good example of an information-rich technology that will help to continue to foster social presence. Additionally, some virtual project managers have created simple social networking websites for their teams that provide member profiles. (More on these technologies in Chapter 8.)

Celebrate Success as a Team

Making sure that the entire team participates in team celebrations goes a long way toward focusing team members on team accomplishments over individual accomplishments. The celebrations do not have to be large or even formal in nature to be appreciated by team members. Successful virtual project managers look for additional opportunities to recognize team accomplishments throughout the duration of the project at key milestones, major events, and even as surprises to the team to help alleviate anxiety and pressures.

Building Trust

Trust must exist between project managers and team members and between team members themselves if a high level of performance is to be achieved. Several factors are important to the establishment and sustainment of trust. These factors include open and honest communication, having no fear of reprimand or reprisal, and confidence that conflicts will be dealt with successfully and properly.

In his book *The 21 Irrefutable Laws of Leadership*, John Maxwell uses the analogy of building trust as either putting change into your pocket or paying it out.[18] Each time project managers form a new team, they begin with a certain amount of change in their pocket, representing the inherent trust a person receives from his or her position as the team leader. As the project progresses, team leaders either continue to accumulate change in their pocket by building trust or find that the pocket begins to empty when trust is depleting.

This analogy holds true for leaders of virtual project teams as well, but there is a distinct difference. Because of the geographic separation and cultural differences inherent in a virtual team, project managers and team members may find their pockets completely empty of change at the beginning of the project—meaning that a virtual team may begin collaborating within an environment that is completely lacking trust. For traditional teams, some trust may exist based solely on social bonds that are in place given their co-location and commonality in culture. Virtual teams do not have this social bond as a foundation of trust to build on. Trust in a virtual environment is granted to those who demonstrate they are trustworthy; therefore, it is based more on consistent and proven performance by both virtual project manager and team members than on social bonds.

Table 3.1 lists the factors that both destroy and create trust on a project team. Obviously, the team leader is best served by acting on the trust creators and avoiding the trust destroyers.

Building strong relationships between team members is also an important factor in enhancing and sustaining trust on virtual project teams, especially later in the team's life. Because of geographic separation, creating and sustaining trust on virtual teams requires a more conscious and planned effort on the part of project managers. At a minimum, it requires managers to spend more time networking and traveling across geographic boundaries.

Establishing Team Norms

A project team can never reach a state of high performance if it does not perform and interact in a consistent manner. Consistency comes from establishing and following a set of team norms that should be established in the early stages of team formation (ideally be documented in the team charter). Norms are the rules and guidelines that a team agrees to follow as it conducts its work.[19]

Table 3.1: Trust Creators and Destroyers

Trust Creators	Trust Destroyers
Act with integrity	Demonstrate inconsistency between words and actions
Communicate openly and honestly	Withhold information or support
Focus the team on shared goals	Put personal gain over team gain
Show respect to team members as equal partners	Listen with a closed mind

On traditional projects, team norms generally are unwritten and engrained in the organizational culture. For virtual projects, establishing team norms needs to be more deliberate. The norms need to be put in writing and to be reviewed on a periodic basis. Once developed, agreed upon, and implemented, team norms begin to guide the behavior and interpersonal interactions of team members and build team discipline, trust between members, and predictability in behavior. These established norms contribute to a solid foundation for the project team; if these norms are not agreed to and implemented, there is a greater probability for potential misunderstanding and conflict between team members.

Virtual project managers will find it beneficial to begin with an initial set of norms. Although team norms are unique to organizations and specific projects, Table 3.2 provides examples that are common for high-performance teams.

The team should also recognize that the team norms are set based on what is known by the team and team leader at the time of project initiation. As the project progresses, team norms may need to be adjusted, and new ones may need to be established and implemented.

Table 3.2: Team Norms

Team Category	Example Norm
Personal behavior	Always treat other team members with professional courtesy and respect.
Communication protocols	For voice communication the team will use Skype technology. For text messaging, the team will use Skype or Windows Live Messenger.
Commitment	Set and adhere to commitments made and agreed to by the due date, assisting other team members when requested, and don't surprise the project manager with missed commitments.
Conflicts	Conflicts will be identified and discussed with team members involved. Viable alternatives and solutions will be developed and discussed. Conflicts will be resolved in a timely, positive, and constructive manner
Meetings	Meetings will begin on time. There will be a set and prepublished agenda. Meeting notes will be created and distributed.
Decision making	Decisions will be made in a consultative manner with one decision maker appointed.
Consent	Disagreements must be voiced and discussed, silence means consent.
Respect	No put-downs, either in foreground or background discussions, will be directed toward fellow team members.
Timeliness	Voicemails and emails must be returned within 24 hours and team members will own the timeliness and quality of our work.
Bias toward action	Team members will lend support, raise concerns, and praise great work in a proactive manner.
Celebrate success	Both large and small successes will be celebrated to recognize accomplishments.

Empowering the Team

With leadership comes power. The most effective project managers and team leaders are those who are willing to share their power with those team members who can make the most positive impact. As trust develops on a project team, project managers should begin to delegate some power to other key team leaders by empowering them. Empowerment is the sharing of power from one person to another and granting others influence and authority to take responsibility and make decisions within their sphere of work. As explained previously, team empowerment means giving the project team members the responsibility and authority to make decisions at the local level. In their book *The Power of Product Platforms*, Marc Meyer and Alvin Lehnerd state: "There is no organizational sin more demoralizing to teams than lack of empowerment."[20]

As team members are granted greater power, they will begin to act more independently and rely less on the direction of the team leader. They will take on a greater sense of responsibility for their work output, become more comfortable with making decisions and solving problems on their own, begin to act proactively instead of reactively, and ultimately become more motivated to succeed. Empowerment of virtual team members motivates and encourages them to work, make decisions, solve specific problems facing their team assignments, and take the necessary actions autonomously through self-management.[21] Without empowerment to decide and act, virtual team members simply wait, usually in frustration, until decisions are made and passed down to them. This waiting is inefficient, ineffective, and demoralizing.

Team member empowerment is arguably more critical on virtual projects than on traditional ones as it is an effective tool for quick and effective decision making, where people closest to an issue are most suited and able to evaluate the situation and decide the proper course of action. Greater discussion on team empowerment is provided in Chapter 6.

Selecting the Right Team Leader

For virtual teams to operate at a high level of performance and ultimately succeed in their mission, strong leadership from project managers is a must. Although the skills and abilities project team leaders need for traditional projects are similar to those needed for virtual projects, there are a number of differences due to the fact that virtual teams have challenges caused by differences in location, time zones, and culture. The key to overcoming these challenges rests in large part on the shoulders of project managers as they fill the role of team leader.

Senior managers must do their best to select project managers who have the ability and the experience to build a high-performance team. When selecting a leader, effort should be made to select people with the ability to balance both the execution-oriented practices and the leadership practices that define virtual project work. More than anything, these all-too-common practices should be avoided: selecting the first person to volunteer for the position; selecting one who has success leading only traditional teams; selecting managers based only on certification, "entitled experts," or the best technical skills.[22]

Selecting the right team leader for a virtual project begins with understanding key characteristics that effective virtual project team leaders possess. Listed next are core characteristics that should be sought out when considering individuals to take on the leadership role of a virtual project team. Team leaders should:

- Be comfortable working in an unstructured and at times ambiguous environment.
- Motivate others to achieve results.
- Delegate work and responsibilities effectively.
- Provide effective governance without micromanaging individuals.
- Demonstrate strong communication, provide clear direction, and are responsive.
- Effectively manage conflict.
- Recognize and rewards others.

Most important, effective team leaders inspire personnel to collaborate as a team regardless if they are co-located or distributed. They inspire and motivate their teams to achieve not only the best level of individual performance but also to relegate their individual performance to that of greater team performance. They do this by communicating higher expectations of the team through the efforts of the individuals. Effective and inspirational communication transcends time, distance, and cultural barriers.

Adjusting Team Membership

When asked about the team member selection process for his virtual project teams, Kirk Rheinhold, the assigned project manager, presented a situation that is common in many organizations. "I have little influence on who is actually assigned to my project team, especially on projects that have team members from multiple company sites," he explained. With virtual teams, especially teams that are widely distributed or consist of members from multiple companies, virtual project managers rarely have influence over who is initially assigned to work on their project.

The difficulty with this reality of modern project life is that the ability to build a high-performing team is highly dependent on who is on the team, their personalities, how well they work as team members, and their work ethic. When a virtual project team is formed by people other than the project manager, these performance dependencies are unknown.

This does not mean, however, that project managers are powerless, as Rheinhold explains: "I go into a new team with the base assumption that some level of adjustment to team membership will be needed. The project manager must be able to quickly assess the abilities of the team members to effectively work in a virtual team environment. Then if problems exist, they must take it upon themselves to drive for change in membership or work with line managers to have the necessary training, mentoring, and coaching provided."

Personal attributes needed for a person to work well in a virtual project environment are normally not well understood within an organization. It is just assumed that *anyone* can work virtually. As a result, project team members are seldom screened for virtual attributes during the team member selection process. In particular, virtual team members must be considerably more self-sufficient than their counterparts on traditional project teams who can more easily seek help in their team environment. Virtual team members also must be able to tolerate ambiguity better than members of traditional teams due to the additional lack of clarity and other complexities caused by the time and distance issues associated with highly distributed teams.[23]

Virtual team members must also be able to keep their emotions in check. Emotionally charged communications lack context and, as a result, often become a source of miscommunication and conflict in the virtual environment. In such environments, emotions should be employed to enable and improve reasoning during problem solving and decision making and to positively affect the emotions of others.

Virtual project teams rely a great deal on technology to communicate, collaborate, and perform their work. Therefore, technological sophistication is required of all team members. Members must be willing and able to adapt to and use new technologies meant to enhance team communication and collaboration.

Finally, virtual project team members must always be willing to share information with their fellow team members. Transparency of information is a keystone factor in raising a team's performance level team. The distributed team model provides ample opportunity for information to be kept local, so the team members themselves must be motivated to ensure that their information is widely distributed and available to other team members and stakeholders.

These are a few of the critical personal attributes to look for in virtual team members, but there are others that virtual project managers can identify as necessary for their teams. The burden of assessing team members' ability to perform on a virtual team and of making personnel changes when necessary rests with project managers. An assessment tool is provided to assist virtual project managers in assessing team members.

Assessing Virtual Project Team Members

The Virtual Project Team Member Assessment is used by organizational management and virtual project managers to evaluate and select team members to serve on virtual project teams. The assessment is divided into two sets of criteria. The first set of criteria, Basic Criteria, represent capabilities that are basic to all teams, both co-located and virtual. The second set of criteria, Virtual Team Criteria, is intended to assess the capability and experience level of potential team members.

Each possible virtual project team member can be evaluated separately. To evaluate each candidate, project managers and other leaders should compare assessment results with one another. People should be offered virtual project team member positions based on their individual potential for successfully achieving the project objectives and senior management expectations for the project. The sample assessment included can serve as a baseline for a customized assessment that fits specific organizational needs.

Virtual Project Team Member Assessment

Date of Assessment: _____

Candidate Team Member Name: _____

Virtual Project Team Leader Name: _____

Assessment Completed by: _____

Estimated Size of Team: _____   Number of Team Sites: _____

Confidential Assessment: _____Yes, confidential
                          _____ No, not confidential

Assessment Item	Yes or No	Notes for All No Responses
Basic Criteria		
The candidate has proven technical or specialist skills to contribute to the project team.		
The candidate has proven project management skills.		
The candidate has proven problem-solving skills.		
The candidate has successfully participated on or led co-located project teams.		
The candidate has demonstrated self-motivation and ability to perform self-directed work.		
The candidate has a track record of ethical behavior.		
The candidate has proven to be reliable and dependable.		

Assessment Item	Yes or No	Notes for All No Responses
The candidate has demonstrated personal confidence in his or her work.		
The candidate has a track record of trustworthiness.		
The candidate demonstrates a positive attitude.		
The candidate has a proven track record of meeting his or her commitments.		
The candidate has demonstrated the willingness to share project information.		
Virtual Project Team Criteria		
The candidate has prior experience with virtual teams.		
The candidate has prior international experience.		
The candidate has proven ability to deal with ambiguity.		
The candidate has proven networking skills.		
The candidate communicates concisely and clearly.		
The candidate is sufficiently proficient in the language of the project.		
The candidate is proficient in other languages besides the language of the project.		
The candidate has proven proficiency with electronic communication technology.		
The candidate has proven ability to work independently.		
The candidate has the ability to cope with isolation from rest of team.		

Findings, Key Thoughts, and Recommendations

Notes

1. Parviz F. Rad and Ginger Levin, *Achieving Project Management Success Using Virtual Teams* (Plantation, FL: J. Ross, 2003).

2. Rad and Levin, *Achieving Project Management Success.*

3. J. L. Creighton, "Team Formation in Remote Virtual Teams." New Ways of Working Network, New WoW Project of the Built Environment PRE Program, August 2011. Rym Oy, Finland. http://NewWoW.net/Sites.

4. *Business Dictionary* website, http://www .BusinessDictionary.com.

5. Creighton, "Team Formation in Remote Virtual Teams."

6. J. R. Katzenbach and D. K. Smith, *The Wisdom of Teams: Creating the High Performance Organization* (New York, NY: HarperCollins, 2003).

7. Katzenbach and Smith, *The Wisdom of Teams.*

8. P. Rogers, P. Meehan, and S. Tanner, "Building a Winning Culture," Bain and Company, August 25,

2006. http://www.bain.com/publications/articles/building-winning-culture.aspx.

9. B. W. Tuckman, "Developmental Sequence in Small Groups," *Psychological Bulletin* 63, no.6 (1965): 384–399.

10. B. W. Tuckman and M. C. Jensen, "Stages of Small Group Development Revisited," *Group Organization Management Journal* 2, no. 4 (December 1997): 419–427.

11. Creighton, "Team Formation in Remote Virtual Teams."

12. Russ J. Martinelli, James Waddell, and Tim Rahschulte, *Program Management for Improved Business Results*, 2nd ed. (Hoboken, NJ: John Wiley & Sons, 2014).

13. Jim Waddell, Tim Rahschulte, and Russ J. Martinelli, "Effective Global Team Leadership Practices." *PM World Today*, no. 11 (November 2010).

14. Martinelli and Milosevic, *The Project Management ToolBox*.

15. Russ J. Martinelli and Dragan Z. Milosevic, *The Project Management ToolBox*, 2nd ed. (Hoboken, NJ: John Wiley & Sons, 2016).

16. M. Beck et al., "Working in a Virtual World: Establishing Highly Effective Virtual Teams on Information Technology Projects." Deloitte White Paper, 2011. http://www2.deloitte.com/content/dam/Deloitte/mx/Documents/human-capital/Working_Virtual_World.pdf.

17. D. DeRosa and R. Lepsinger, *Virtual Team Success: A Practical Guide for Working and Learning from a Distance* (San Francisco, CA: Jossey-Bass, 2010).

18. John C. Maxwell, *The 21 Irrefutable Laws of Leadership* (Nashville, TN: Thomas Nelson, 1998).

19. Donald J. Bodwell, "High Performance Teams: Team Norms." www.highperformanceteams.org/hpt_norm.htm.

20. Marc H. Meyer and Alvin P. Lehnerd. *The Power of Product Platforms* (New York, NY: Free Press, 1997).

21. M. R. Lee, *Leading Virtual Project Teams: Adapting Leadership Theories and Communications Techniques to 21st Century Organizations* (Boca Raton, FL: CRC Press, 2014).

22. *Business Week Magazine* (June 2009).

23. K. Ferrazzi, "To Make Virtual Teams Succeed, Pick the Right Players," *Harvard Business Review*, December 18, 2013. https://hbr.org/2013/12/to-make-virtual-teams-succeed-pick-the-right-players.

MANAGING PROJECT EXECUTION, LEADING THE VIRTUAL TEAM

EXECUTING THE VIRTUAL PROJECT

The project manager's planning work is just that, planning. There is a reason you see coaches and managers on the sidelines during games; they are needed as much then as they are needed during practice. If practice was the only important aspect of a coach's role, there would be no need for her or him during the game, where all that practice is executed. The same can be said for virtual project managers as projects move into execution.

If planning work was done well, an integrated project plan will be created that represents the collective work that needs to be accomplished by the team during project execution. If best practices for virtual projects were used, there was likely one or more face-to-face meetings to solidify the project charter and team charter, cement project member roles and expectations based on the project's business case, demystify any assumptions or rumors about any team member and cultural beliefs, and enjoy some social time to get to know one another and establish team chemistry. Hopefully, this did occur; even if it did, though, it does not mean that the synergy will be maintained once the team and the work are redistributed geographically, organizationally, and culturally. A one-time face-to-face meeting or event does not make a team perform at a high level. The most important aspect of any such event is not the event itself but what happens afterward. (See the box titled "Actions Speak Louder than Words.")

Actions Speak Louder than Words

Recently, one of the authors was involved in an organizational acquisition. A global organization headquartered on the East coast of the United States acquired our (largely) co-located and much smaller organization in the Pacific Northwest. Time will tell if the acquisition is successful. On paper, it was a great deal for both organizations with good strategic fit, little redundancy in customer base and product offerings yet good synergy in mission and focus, and both organizations are financially strong.

At the time of the acquisition, best practices were observed. There was good (clear, concise, and confident) communication with an equally good post acquisition integration plan. One of the first activities was to get the integration teams together for a retreat. There was a three-day meeting of a few dozen people at the headquarters of the acquiring firm.

The agenda for the integration planning retreat was full yet balanced with acquisition commentary from the CEO, integration expectations from the CFO and COO, and social time to connect the teams to build trust and rapport.

As the three days concluded, people spoke of time well spent and gratitude for new connections made from opposite sides of the country. And there was optimism that reinforced the strategic reason and good fit of the acquisition. Everyone felt they had what they needed to succeed, and there were commitments by all to consistently communicate status and collaboratively resolve issues.

As can be imagined, participants in the three-day integration planning retreat were met with a number of questions from coworkers and colleagues back at the home office. Many did not have clear answers to those questions. There was a general email address that could be used for questions, but most that were submitted went unanswered. The following week, members of the multiple workgroups of the post acquisition integration team realized that they did not have a good way of communicating across the teams and there was no common registry of who was on what team. Some members quickly found themselves allocated over 200% of time to various workgroups—that was in addition to their already 110% allocation to their regular job.

A snapshot taken a month into the post acquisition integration team's efforts showed many projects already behind schedule. The schedule being used was the CFO's cost savings schedule along with the COO's product integration expectations, neither of which were collaboratively created. The schedules looked good on paper, but the breakdown in execution was due in part to the distributed nature of the teams. Many team members at the distributed sites simply didn't know what they needed to accomplish.

At the retreat, there was no time set aside to discuss project expectations, communication plans, or cadence of status reports and dashboards. Many of the 21 projects were reset and recast after the retreat due to lack of communication, organization, monitoring, and control. This had negative implications for employees, especially those not directly involved in the post acquisition projects. Skeptics and naysayers started to voice their opinions, which cascaded through the organizations. The project that started off with best practices was now struggling.

The lack of social interaction and communication led to diminished trust and loss of team spirit that was present during the retreat. Now the signs of virtual project killers were evident. When the best intentions to communicate are not followed up, project teams suffer. Best intentions are never enough. It's true what they say: Actions speak louder than words.

The executing stage is arguably the most anticipated stage of a project's life cycle. It is when an intangible concept moves forward to become a usable, tangible asset for the business. It is also the part of the project in which the quality of the integrated project plan and the ability of project managers will be put to the test.

The job of managing a virtual project during the executing phase can be challenging due to a couple of natural factors. First, the size of the project team and the number of project interdependencies to track and manage grow rapidly from planning to executing. Staff size, budgeted dollars, and the number of project interdependencies are at their highest levels during this stage of the project. Project managers must closely manage these factors to ensure that the project stays in alignment with the business objectives and that the triple constraints are balanced.

The second factor that complicates the management of virtual projects is, of course, the geographic distribution of the work caused by the distribution of the team. Monitoring, adjusting, integrating, and changing the work of the project team is complicated by the distance and time factors associated with virtual projects. Because of these complex factors, there are more nuances for virtual project managers to address. This chapter details those nuances, best practices, and solutions to enable project managers to overcome the challenge of virtual project execution.

Managing Assumptions

Few, if any, projects begin with absolute certainty of how their execution and outcome will play out. In reality, more is unknown than known during the planning stage of a project. This is due to the fact that project planning is about trying to predict the future, and even the most gifted project managers do not possess a crystal ball to foresee future events. As projects are planned and executed, some facts are known while the remainder must be estimated based on a set of assumptions of how the future will unfold. An *assumption* is a likely condition, circumstance, or event that is presumed known and true.

Assumptions and constraints form the basis of the project plan and must be closely monitored during project execution to ensure that the plan remains viable. If not, changes will need to be made. For example, a constraint for a particular project may be that $200,000 has been allocated to complete the project (the project budget). A base assumption of the project may be that the $200,000 budget is sufficient to complete the project. If this assumption is proven untrue, changes will need to take place during project execution to realign reality and the constraint.

Project Assumptions Identification

The first and most crucial step in managing assumptions is identifying the base assumptions on which the project plan is established. If assumptions are not identified and made visible, they cannot be managed. Identification of assumptions is important on all projects, but even more so for virtual projects because the validity of the assumptions has to be assessed wherever the work is being performed. Managing the base assumptions is therefore a shared and distributed task on virtual projects. For this reason, it is necessary to clearly document the base assumptions for the project, as illustrated in Table 4.1.

Assumptions Validation

The best-case scenario is one in which all assumptions are validated as truisms during project planning so the team begins execution activities with a full set of knowns. This is an extremely rare scenario, unfortunately. In reality, the validation of assumptions becomes a primary activity during project execution that feeds both the risk management and change management processes. Because of its importance, assumption validation has to be centralized and performed by virtual project managers (with critical input from project team members). As a best practice, the assumption list should be reviewed during each team meeting to maintain focus and effort on validating the base assumptions of the project.

Assumptions Control

Initial assumptions are rarely static. As a project evolves, more will be learned about the future,

Table 4.1: Base Assumptions for Project "Amherst"	
Assumption	**Status**
Resources will be available at project kick-off.	Open
Technical staff is fully trained.	Confirmed
Business requirements are completely documented.	Open
Full budget is available at project initiation.	Confirmed
Equipment order lead times are known and can be met.	Open
System components will be able to be integrated with minimal rework (three weeks maximum).	Open
No outsourced work will be required.	Confirmed
Strategic customers will participate in beta testing of the product.	Open

and assumptions will be proven true or untrue in the process. When an assumption is proven to be untrue, action must be taken. Likely, a project risk will be associated with the false assumption that will require mitigation or elimination. The risk mitigation or elimination action will in turn likely result in a change to the project baseline. Project assumptions, risks, and changes need to be closely managed as they, many times, are closely related and dependent on one another.

Documenting and validating the base assumptions on which the project plan is established enables a more proactive approach to managing the execution of a virtual project. Risks and changes can be identified and responded to in a timelier, and many times in a less disruptive manner if assumptions are tested early and regularly as part of the project execution process.

Hyper-vigilant Governance

"Monitoring progress on a virtual project is like status on steroids." That is how Giovanni Mazzolli sees governance for a virtual project. Mazzolli, a virtual project manager in the automotive industry, is not alone in his view of virtual project governance. When we began our research for this book, we established a panel of experienced virtual project managers. The goal in doing so was to get a wide variety of practice-based perspectives pertaining to managing virtual projects and leading virtual teams. We began with one question: "What are the three main differences between traditional projects and virtual projects?" We received many great answers, most of which are included in this book. However, the answer we received the most often supports Mazzolli's assertion that hypervigilant project governance is a primary discriminator between traditional and virtual projects. Here is a sample set of other responses we received:

■ "I have experience with both co-located and virtual teams and I find the main difference is during execution and when performing monitoring and control."

■ "When you track progress during execution, I find it very different when it has to be done remotely. When done remotely, the process of reporting is more structured. We set the days we meet and the days I get an email status report."

■ "When working with a local team, monitoring is less formal. I tour the office every morning and do a 'what's new' survey of team progress. When monitoring remotely, it is more important to use defined reporting cycles and common dashboards."

The distribution of work in the virtual environment makes project monitoring and control so challenging and so important. Studies regarding virtual team governance reveal that stage gate–based systems are reasonably standard in project governance for reviewing and approving established project milestones and critical decision points. Further, metrics and dashboards are used for measuring and tracking progress and related success factors and are used as decision support tools. Such tools are used to keep execution aligned with business objectives and to illustrate what is working well and what needs additional effort relative to the project success factors. The governance process and tools will indicate when course correction becomes necessary.[1]

A solid governance system also requires that periodic reporting and assessing takes place consistently and across the organization by each project manager and project team. Additionally, responsibility and authority is assigned to managers and project managers possessing governance responsibilities to ensure that actions and decision making are occurring as intended. The consistent application of the governance system is necessary to provide the highest probability of achieving the intended project objectives. This point cannot be overemphasized as it pertains specifically to virtual projects that face challenges associated with the virtual team environment.[2]

The Project Management Institute defines *project governance* as the alignment of project objectives with the strategy of the larger organization by the project sponsor and the project team.[3] A project's governance is defined by and is required to fit within the larger context of the organization sponsoring it, but is separate from organizational governance. Project governance manifests itself as an oversight activity that most organizations consider a subset of the broader organizational governance activities defined by the project life cycle. The project governance framework normally includes such things as structure, processes, decision making models, and tools for monitoring and controlling the project. A robust project governance system has three primary functions:

1. Establish and maintain the project goals, objectives, and linkage to the organization's strategic goals.

2. Ensure the right structures are in place to achieve the project goals and objectives.

3. Monitor and direct the project to ensure that stated project objectives and business benefits are realized.

Most organizations have a strong link between internal project governance activities normally managed and led by project managers and the broader organizational governance system established by the enterprise. Projects are initiated to achieve business goals, and as a result, project success is judged on the basis of the achievement of those goals.

Start with Defining Project Success

The first principle of good project governance involves understanding and communicating how well project performance is progressing toward achievement of project objectives. Many times project managers become over focused on progress against their cost and schedule baselines and forget that the real intent of a project is to achieve the business objectives driving the need for the project.

The project strike zone is an excellent tool for evaluating and communicating progress toward achievement of the project objectives. It is used to identify the critical objectives for a project, to

help project managers and their stakeholders track progress toward achievement of the key business results anticipated, and to set the boundaries within which project managers and teams can operate without direct top management involvement.

As shown in Figure 4.1, elements of the project strike zone include the project objectives, target, and threshold values, an "actual" field that provides indication of where a project is operating with respect to the target and threshold limits, and a high-level status indicator.[4]

Creating an effective project strike zone is a critical activity for ensuring that project managers, project teams, top management, and other stakeholders all understand and agree on the objectives of the project. It is also critical for establishing the success measures to monitor during project execution in order to gauge project progress.

Defining a meaningful project strike zone requires quality information from a number of sources. The initial set of objectives is derived directly from the approved project business case. (See Chapter 2.) To establish and later negotiate the control limits for each objective with project managers, project sponsors also need to know the project team's capabilities and experience and past track record and to balance thresholds against the new project's complexities and risks accordingly. The four steps for creating and using a project strike zone are:

1. Identify project objectives.

2. Set the recommended target and threshold values.

3. Negotiate the final target and threshold values.

4. Use the project strike zone for virtual governance.

Identify Project Objectives

Identification of the project objectives begins during the initiation stage of a project. The objectives represent a subset of the metrics normally tracked by a project team. The project strike zone should include only the measures that represent the high-level

Project Strike Zone

Project Objectives	Strike Zone		Actual	Status
Value Proposition	**Target**	**Threshold**		
• Increase market share in product segment				
• Order growth within 6 months of launch	10%	5%	7% (est.)	Green
• Market share increase after 1 year	5%	0%	4% (est.)	
Time-to-Benefits Target:				
• Project initiation approval	1/3/2019	1/15/2019	1/4/2019	
• Business case approval	6/1/2019	6/30/2019	6/1/2019	
• Integrated plan approval	8/6/2019	8/20/2019	8/17/2019	Red
• Validation release	4/15/2020	4/30/2020	6/29/2020	
• Release to customers	7/15/2020	8/1/2020	TBD	
Resources				
• Team staffing commitments complete	6/30/2019	7/15/2019	7/1/2019	
• Staffing gaps	All project teams staffed at minimum level	No critical path resource gaps	Staffed	Green
Technology				
• Technology identification complete	4/30/2019	5/15/2019	4/28/2019	
• Core technology development complete	Priority 1 & 2 techs delivered @ Alpha	Priority 1 techs delivered @ Alpha	on track	Green
Financials				
• Program budget	100% of plan	105% of plan	101% (est.)	
• Product cost	$8500	$8900	$9100 (est.)	Yellow
• Profitability index	2.0	1.8	1.9 (est.)	

Figure 4.1: Example Project Strike Zone

project objectives (often the business objectives). The project objectives will be unique to every organization and are derived directly from strategic management and portfolio management processes.

The strike zone is most effective when the objectives identified are kept to a critical few (usually five to six), as this focuses the project and top management's attention on the highest priority contributors to project success. The factors deemed must-haves often include market, financial, and schedule targets and the value proposition of the project output.

Set the Recommended Target and Threshold Values

The target and threshold control limits shown in Figure 4.1 form the strike zone of success for each project objective. The target values for the objectives in the project strike zone are the objectives as specified in the project business case and baseline plan. The target values should be pulled directly from the project business case.

The threshold values represent the upper or lower limit of success, as specified by senior management for the project objectives.[5] Some discussion and debate is normally required to understand how far off target an objective can range and still constitute success for the project. For example, the target project budget may be set to $500,000. But, if additional spending of 5% is allowable, then the budget threshold can be set at $525,000. This means that even though a project team misses the target budget of $500,000, it still is successful from a project budget perspective if it spends up to $525,000.

Negotiate the Final Target and Threshold Values

Once project managers establish the recommended target and threshold values for each project objective, they present the information to the senior executive sponsoring the project. Based on the project's complexity and risk level, and on the capability and track record of the project manager and team, the project sponsor may adjust the values accordingly. For example, on a project that is low complexity, low risk, and is being managed by an experienced project manager, the range between target and threshold values may be opened up to allow for a higher degree of decision making empowerment for the project manager. Conversely, on a project that is of higher complexity or risk or is being managed by an inexperienced project manager, the range between target and threshold values will be tightened up to limit the project manager's decision making empowerment.

Once the targets and boundaries are negotiated, the team should be empowered to move rapidly as long as members do not violate the strike zone threshold values.

Use the Project Strike Zone for Virtual Governance

The project strike zone adds value in many ways to project managers, the executive sponsor of a project, and the project governance body. Project managers utilize it to formalize the critical project objectives for the project, to negotiate and establish the team's empowerment boundaries with executive management, to communicate overall project progress and success, and to facilitate various trade-off decisions throughout the project life cycle.

Executive managers utilize the project strike zone to ensure a project supports the intended business objectives and to establish control limits to ensure that the project team's capabilities are in balance with the specific project's level of complexity. When used properly, the project strike zone provides top managers a forward-looking view of project alignment to business objectives. When problems are encountered, the tool's structure provides an early warning of trending problems, followed by a clear identification of "showstopper" conditions based on the level of achievement of project objectives. If a project is halted, senior executives can reset the project objective targets or thresholds, modify the scope of the project to bring it within the current targets, or, in the extreme case, cancel the project to prevent further investment of resources.

Executive managers and the project governance body set the boundary conditions (targets and thresholds) of the project strike zone between which project managers can operate, thereby empowering project managers to make decisions and manage the project without direct top management involvement. As long as the project progresses within the strike zone of each project objective, the project is considered on track, and project managers remain fully empowered to manage the project through its life cycle. However, if the project does not progress within the strike zone of each project objective, the project is not considered on track, and the top managers must intervene directly.

When the project strike zone is used appropriately, the project manager and team are empowered because boundaries for authority, responsibility, and accountability are established. Too often, project managers tell us that they have all the responsibility for driving project success but lack the authority. The project strike zone is the best tool we are familiar with to balance responsibility with the appropriate level of authority.

Monitor Distributed Work Closely

Project monitoring is a critical aspect of virtual project governance. Due to the distributed nature of work on a virtual project, project managers have to be hyper-vigilant about collecting team status and monitoring progress to plan. They must do so on a schedule specified by their respective senior management and project sponsor in an organized manner.

Many times monitoring cadence is dependent on where the project is in its life cycle, as Jerry Conners, an experienced virtual project manager, explains:

> During the early and middle stages of project execution, I rely on our weekly team meeting to collect status from the distributed team members. This is my normal reporting cycle. During final integration and testing, we will go to daily stand-up meetings to make sure action and re-action is quick.

Monitoring is the delicate balance between not diving too deep, because virtual project managers will quickly realize that there is not the time to do so, and not diving deep enough to monitor project progress and team health. Best-practicing virtual project managers realize that as managers, they are part auditor and therefore need to take on a "just don't tell me, show me" attitude (see the box titled "Using Weekly One-on-Ones to Capture Dashboard Data and Build Rapport").

Project managers need to see work completion to fully trust it is getting done. It is a trust but verify management style and is accomplished in two ways. The project managers intentionally design project schedules and integration points that are closely associated in time. In other words, best-practice virtual project managers never have decision points, milestones, or deliverables that are months into the future. All such key points on a schedule are no more than two to three weeks out, which means there are always touchpoints to verify actual progress.

Using Weekly One-on-Ones to Capture Dashboard Data and Build Rapport

Virtual project managers know the importance of meetings. The best virtual project managers never pass up an opportunity to have a meeting. That may sound odd, but these project managers know the value of a good meeting, not for the sake of the meeting, but for team cohesion.

To be successful, project managers have to be visible. This means they must regularly, routinely, and intentionally connect with team members and stakeholders. To make sure there are no surprises and risks are being managed, great project managers know the importance of weekly one-on-one meetings with each of their team members. Depending on what is going on that week, the duration of each meeting will vary. Some may be 15 minutes, but others may be an hour. The goal of each meeting is fourfold:

1. Ensure data are captured to track on the dashboard.

2. Make, create, and sustain a positive personal connection.

3. Make sure the team member knows the value of her or his contribution to the team, project, and organization.

4. Make sure the team member has the resources to be successful.

Similar meetings should occur with key stakeholders, not the least of whom would be sponsors and steering committee members. Project managers should plan accordingly and not miss an opportunity to make a meaningful connection in their next one-on-one.

Software Development Team Indicator

Software Development Status

Work accomplished last week:
- Updated latest software build with User Guide

Work planned for this week:
- Debug critical bugs
- Complete Linux support plan

Next deliverables:
- Linux support plan –due 10/14
- Software build 42 – due 11/01

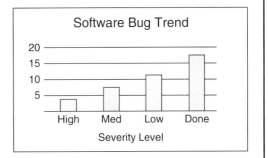

Software Bug Trend

Critical Bugs

1. Operating system is crashing
2. New firmware release causes system interrupts

Risks

Risk: Linux developer not rolling off Icon program when expected

Impact
Schedule will be delayed 2 weeks

Mitigation
Borrow developer from technical marketing team for short-term relief

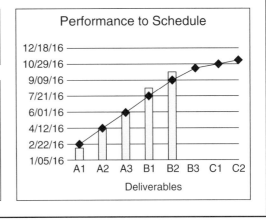

Performance to Schedule

Figure 4.2: Example Team Indicator Report

To facilitate consistent intra-team progress reporting, virtual project managers can require team leaders to prepare and present a team indicator that reflects the work of their team of specialists. Keep in mind that, for some distributed team members, it may be a team of one specialist. Figure 4.2 shows an example project team indicator report that can be adapted for a team's special situation and information requirements.

Project managers must work with the team to determine the best format and content to present in the team indicator. Content includes progress to planned work since the *last* reporting period, work planned during the *next* reporting period, critical issues and risks needing project manager attention, changes agreed to through the team's change management process, and performance to a team's performance metrics. By requiring team members to keep their indicators current, project managers will receive a comprehensive yet concise report on team status. As an effective communication tool, the team indicator facilitates the necessary transfer of status information between the distributed team member and the virtual project manager.

Consolidating Project Status Information

In today's frenzied pace of virtual projects, project managers need to understand how the project they are responsible for is performing, but they rarely have time to read through a number of detailed status reports from their functional teams. From this time-versus-information dilemma grew the concept of the project dashboard.[6]

Much as the dashboard of an automobile provides drivers with a quick snapshot of the current performance of the vehicle, the project dashboard provides project managers with an up-to-date view of the current status of the project. Unlike the project strike zone, which focuses on performance against the higher-level project objectives and business goals, the project dashboard focuses on the current state of the lower-level key performance indicators (KPIs).

The dashboard should be designed as an easy-to-read and concise (often a single page) representation of all KPIs as illustrated in Figure 4.3. The information presented on a dashboard will vary by design, but typical items include:

■ Key decision dates (this week, next 3 weeks, following 2 months)

■ Upcoming milestones (this week, next 3 weeks, following 2 months)

■ Upcoming deliverables (this week, next 3 weeks, following 2 months)

■ All key integration points

■ Project quality

■ Project cost

■ Project risks (top 3)

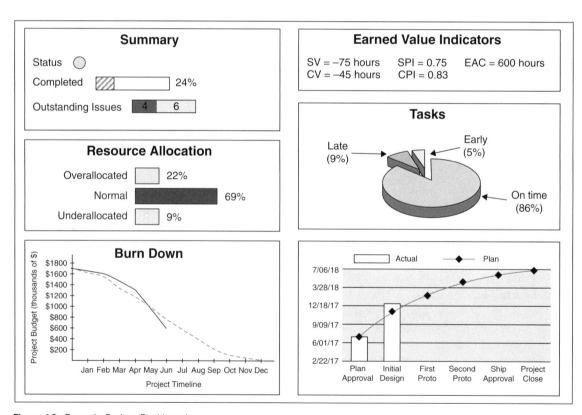

Figure 4.3: Example Project Dashboard

- Key stakeholder engagement meetings/discussions
- Team resourcing metrics (current and forecasted allocation)

There are many types of project dashboards in use and available for reference when designing a customized dashboard that represents the information most relevant and critical to a project. We like the design of the dashboard shown in Figure 4.3 because of its graphical nature, the variety of project status measures it provides, and its conciseness.

Designing a Project Dashboard

The project dashboard is one of the most flexible and customizable tools in a project manager's toolbox. As stated earlier, it needs to be designed around the particular KPIs of a project. Since each project is unique, each project will have somewhat unique performance indicators and therefore likely will have a unique project dashboard design.

Design of the project dashboard begins with the identification of the KPIs for the project. The project objectives, identified and quantified in the project strike zone, define the end-state of the project in terms of what value the project brings to the sponsoring organization. The KPIs quantifiably measure how well the project is performing toward accomplishing the project objectives. These measures typically can be found in other tools, such as the project business case or project charter. (See Chapter 2.)

The project KPIs are part of a measurement hierarchy that must be understood. Business outcomes support an organization's strategic goals, project objectives support the business outcomes, and KPIs support the project objectives. If, for instance, a strategic goal for an enterprise is to be the leader in a particular market segment, a business outcome in support of that strategic goal would be first-to-market advantage with new products or offerings. A project objective would, in turn, have to quantitatively define the project completion date that ensures first-to-market position for the project outcome. Two important project KPIs would likely

complete the measurement hierarchy: performance to schedule and resource allocation percentage. (If resources are not close to 100% allocated to plan, the schedule likely will suffer.)

The KPIs identified in the project dashboard should directly measure performance toward achieving the project objectives documented in the project strike zone. The KPIs represented in the project dashboard in Figure 4.3 include performance to schedule, performance to budget, performance to cost, and resource utilization. Since resources are geographically distributed on a virtual project, resource utilization should always be viewed as a must-have performance indicator for any virtual project.

The final step in designing the project dashboard involves locating the pertinent performance data and representing it on the dashboard. Whenever possible, graphical representations should be used, as they facilitate speedier analysis of the current performance on the part of recipients than text-based representations, and they are easier to communicate to project stakeholders.

Some project managers embed hyperlinks within the top-level performance graphics that link to detailed data about the KPI of interest. For instance, if additional detail is needed for the performance to schedule KPI, a link can be provided to a detailed Gantt chart, milestone analysis chart, or even the schedule section of the current detailed status report for the project.

Project managers can use project dashboards as both communication tools and decision support tools. By using project dashboards to synthesize lower-level performance data into higher-level information, project managers are armed with the information they need to communicate the current status of the project with respect to the KPIs. Additionally, many decisions have to be made during the course of a project, some large and some small, and project dashboards serve as bases of past and current information from which decisions can be driven. (See the box titled "Tips for Using Project Dashboards.")

The overall simplicity of project dashboards may lead project managers to think they are solutions to all problems. Dashboards are only as effective as the design of their structure, the value of the measures and metrics chosen, the accuracy of the data represented, and how effectively they are used to drive communication and decisions.

Project managers must avoid descending into a quantitative and analytical quagmire when using dashboards. The value gained from the use of project dashboards must be greater than the cost of obtaining and analyzing the information contained within the dashboards.

False and conflicting information sometimes may show up in dashboards. Project managers should take the time to ensure that the information is current and accurate and that it conveys the right message about the status of the project against the KPIs. If not, dashboards may do project managers more harm than good.

Project Reviews

The project review is an organizational meeting in which project status is presented to and reviewed by senior managers, governance board personnel, and other key stakeholders. Given the virtual project team environment, most of these reviews are held by video conference across multiple company sites. Each project manager will present the status for the project he or she manages.

For consistency in reporting format and message, it is most effective to have a consistent project summary status report for use by all project managers. Much like the project dashboard, which is used to communicate intra-team status, the project summary status report is a one- or two-page summary brief that communicates overall project status to a broader set of stakeholders. (See Figure 4.4.)[7] Much of the information contained in the status report is derived from the most recent team indicators and presented in summary format.

In addition to the summary status report, the project strike zone and project dashboard tools are normally used in the project review to communicate status toward achievement of the project business goals and KPIs, respectively. It is important to point out that the project review serves a dual purpose. First, it provides the data and information necessary to ensure that a project is progressing as anticipated, and second, it provides the necessary information to assist senior management and the governance body to make decisions with respect to business value and strategic direction. In the project review, the objective for project managers is to communicate both operational and strategic status of the project and to use the forum to gain management help if needed for specific issues or barriers outside of the project managers' control.

Managing Outsourced Project Work

Virtual project teams, just like traditional project teams, periodically face the need to outsource elements of work to other companies or subcontractors. Many times this need arises because the expertise or specialty skills required to complete the work do not reside entirely within the organization or company managing the project.

It is important to note that outsourcing a portion of a project to another company does not relieve project managers of the need to effectively manage the outsourced work to required specifications, delivery, and quality expectations and periodic progress reporting from the owner of the contracted portion of the project.

Leading and managing outsourcing is more challenging for a virtual project team than for a traditional team because of the additional oversight needed due to virtual distribution of the work. Some management and coordination elements for the virtual team leader to keep in mind are listed next.

- Ensure the right virtual project personnel are directly involved in the outsourced work to provide necessary guidance.

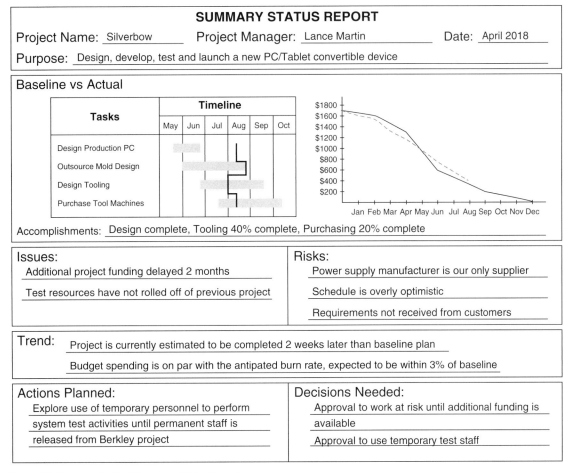

SUMMARY STATUS REPORT

Project Name: Silverbow   Project Manager: Lance Martin   Date: April 2018

Purpose: Design, develop, test and launch a new PC/Tablet convertible device

Baseline vs Actual

Tasks	Timeline					
	May	Jun	Jul	Aug	Sep	Oct
Design Production PC						
Outsource Mold Design						
Design Tooling						
Purchase Tool Machines						

Accomplishments: Design complete, Tooling 40% complete, Purchasing 20% complete

Issues:
Additional project funding delayed 2 months

Test resources have not rolled off of previous project

Risks:
Power supply manufacturer is our only supplier

Schedule is overly optimistic

Requirements not received from customers

Trend: Project is currently estimated to be completed 2 weeks later than baseline plan

Budget spending is on par with the antipated burn rate, expected to be within 3% of baseline

Actions Planned:
Explore use of temporary personnel to perform
system test activities until permanent staff is
released from Berkley project

Decisions Needed:
Approval to work at risk until additional funding is
available

Approval to use temporary test staff

Figure 4.4: Example Project Summary Status Report

- Provide specific and detailed specifications and requirements to the outsourced agent. Most likely, this very specific information will be contained in a written contract.
- Require a project plan from the outsourced agent that specifies schedule, deliverables identified, and testing required before the release of the work. Virtual project managers will incorporate this useful information into the integrated project plan.
- Require periodic status reports and special meetings to address progress, issues, and risks as they arise, and consistently monitor progress of the outsourced work.

- Be ready and available to address the outsourced contractor's questions, concerns, issues, and requests for additional information in a timely manner.

Like all virtual team members, virtual project managers must be willing to invest face time with the outsource agent to build relationships, trust, and a physical presence that will carry forward in time.

Managing Change

Change happens often and quickly on any project. As soon as a project baseline is set, something is sure to change in the project environment that will test the need to adjust that baseline. The same is true on virtual projects, but the probability of unmanaged change creeping in is much higher on a virtual project due to the distribution of work geographically. Diligent and visible management of change is a must on a virtual project and constitutes a large portion of virtual project managers' effort during project execution. To lower the risk of unmanaged change, three practices are important:

1. Gaining visibility to change that is occurring remotely and virtually.

2. Establishing a robust change management process for the project.

3. Efficiently communicating all changes that have been approved and declined to the virtual team and key stakeholders.

Gaining Visibility to Change

The biggest challenge to managing change on a virtual project is dealing with changes that can occur remotely *before they happen*. Proactive management of change can occur only if project managers establish a change management system for their virtual projects. A formal change management system establishes order to the job of collecting, analyzing, documenting, approving, and communicating project changes. Such a system also establishes the protocols, or rules, that define the expectations for how change will be managed once a project is in execution.[8] Change management protocols normally include these points:

- All changes must be requested.

- Change requests must be made in writing.

- The benefits gained from a proposed change must be clearly articulated and documented.

- The approval process must be documented.

- A decision maker must be appointed.

- Approved changes must be incorporated into a revised project plan.

- Change decisions must be adequately communicated.

A formal change management system is useless, however, if the team is not committed to managing change. Project managers must set the expectation that all changes that can potentially affect the project baseline and task plans be communicated and analyzed for cost versus benefit. For distributed teams, it may make sense to establish a change management hierarchy where change is first managed locally, where the work is being performed. Then approved changes are communicated to the project-level change management body for ratification. Only changes that may affect the work of other team members or may affect the project baseline or success criteria need to go to the project level for evaluation. In practice, this is an effective model for collecting and analyzing change that is occurring remotely.

Establishing a Formal Process

The change management process itself is no different for a virtual project than for a traditional project. There is no shortage of materials available to help readers establish a process, so we will not add to the mountain of information. We do want to stress that project managers should strive for simplicity when establishing a change management process with a minimal number of steps. Seven steps are recommended:

1. Submit a written change request.

2. Evaluate benefit versus cost of the change.

3. Assess impact of change.

4. Make the change request decision.

5. Log and communicate the change decision.

6. Modify project and task plans based on approved changes.

7. Communicate and implement the change.

For virtual projects, these steps will be performed with limited to no face-to-face discussions. It is important, therefore, to establish a dedicated change management team site for the project on the company intranet where fast and efficient communication related to various changes can occur. (See the box titled "Fast-Tracking the Change Process.")

Communicating Changes

Finally, change decisions are of no value if they are not communicated to all those affected, especially virtual team members immediately affected by the change. Communication, always a critical factor in virtual project success or failure, is vital to ensure that changes that were approved and declined are communicated to the distributed workforce. Responsibility for implementing changes will likely fall on those performing project execution tasks in remote locations. Regular, specific, and formal change communication must come from the project change management body. The most effective means for communicating change decisions are the use of change management list servers and change management bulletin boards. (See Chapter 8.)

Additionally, communication of change should be incorporated into the project governance process. Each team should include information on changes approved and planned as part of their project indicator. Virtual project managers should, in turn, include baseline changes as part of their project dashboard and summary status report.

Integrating Distributed Work

An inescapable fact of virtual projects is that the various components of a project deliverable and final outcome are designed and developed in different locations.[10] A significant portion of the execution stage of a virtual project involves the integration of the components into a holistic solution. (See Figure 4.5.)

Integration of the work outcomes and deliverables of localized teams is impossible if the work across the various locations of the virtual project is not occurring synchronously. In this regard, project managers are much like orchestra conductors. Even though each of the instrument sections of an orchestra has its own music to produce, the conductor ensures each section steps through the musical composition at a consistent tempo and in concert with each of the other sections. This ensures that an integrated, blended, and harmonious musical piece is produced.

Much is the same on a virtual project except for the fact that the players can be distributed around the world. Since the responsibility for managing the interdependencies between the geographically distributed team members falls on

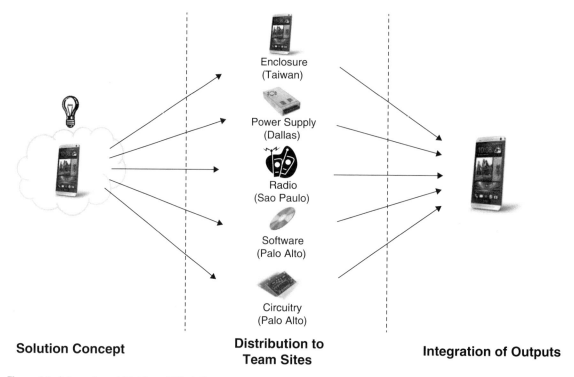

Solution Concept	Distribution to Team Sites	Integration of Outputs

Enclosure (Taiwan)

Power Supply (Dallas)

Radio (Sao Paulo)

Software (Palo Alto)

Circuitry (Palo Alto)

Figure 4.5: Integration of Distributed Work Outcomes

project managers, they must work to ensure that the work of each distributed team member occurs at an integrated and harmonious pace. Synchronization involves ensuring that timelines are aligned, that cross-project deliverables are planned and mapped appropriately, and that the work within each team is occurring at the appropriate pace.

Managing Interdependencies

The disaggregation and reintegration of work is the centerpiece of integration and systems thinking. Virtual project managers oversee the disaggregation of the whole solution into a project architecture composed of multiple components. They then oversee the integration of work output from each of the components to create the consolidated whole solution. The true value and benefits of the project output can be realized only when the activities

associated with each of the team members are integrated together into a holistic solution for the customer.[11]

Referring to Figure 4.6, we consider the horizontal dimension of a project as the synchronization of workflow across the team specialists.

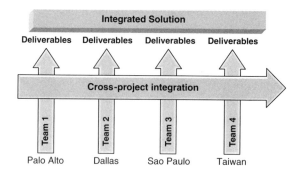

Figure 4.6: Cross-Project Synchronization and Integration

This synchronization is accomplished through cross-project interface definition, coordination of all project activities, collaborative communication on the project, and synchronization of the delivery of the interdependencies.

Although the role of localized teams is to manage the creation and delivery of their deliverables and outcomes, an important aspect of the integration role of virtual project managers is overseeing the handoff of deliverables between team members. This output-input relationship creates a network of interdependencies that has to be established and managed at the project level, not the local level.

Establishing the network of interdependencies is best accomplished through the practice of project mapping (see Chapter 3) and the creation of a project map. Project managers can use the map to manage the hand-off of cross-project interdependencies during the execution stage of the project.

Synchronizing Workflow

For effective integration to occur, synchronization of activities over time is necessary. However, for project managers, time management on a virtual project usually does not just involve managing the duration of tasks associated with each project component. Project managers who attempt to do so would be mired in detail. Rather, managing the project timeline involves ensuring the work outcomes and deliverables are occurring as planned and mapped.

Detailed schedule management is the focus at the local level, and summary, or integrated, time management is the focus at the centralized project level. This coordination requires a modular approach to time management where the schedule is disaggregated and partitioned according to where the work is to be performed. Schedule details for each component are worked by local teams and then integrated by project managers to gain the full perspective of the project timeline and critical synchronization points.

Virtual project managers should keep in mind that the most detailed schedule is not necessarily the best one. Too much detail can divert attention to one aspect of a project—the schedule—and away from the other critical aspects of managing a project and leading a team. If project managers focus on the critical synchronization points occurring over time and let local teams focus on the detailed schedule for their respective project components, a good balance for effective timeline management is achieved.

People Side of Project Execution

The mountain of literature that explains that most projects fail is discouragement enough to stop people from attempting to execute any project. The fact that the failure rate of virtual projects increases in relation to how distributed the virtual team is, puts virtual project managers under even more pressure to maintain diligence on the people side of project execution. Think about it. Here is the work (generally speaking) that project managers oversee during project execution:

- Manage planning assumptions while carrying out the integrated project plan to the realization of the value detailed in the business case and to the specifications in the scope document.

- Manage (in-sourced and outsourced) personnel, task assignments, deliverable hand-offs, and overall performance through tracking and monitoring systems.

- Establish and manage a rigorous governance system and facilitate status meetings and decisions ranging from staffing to task delivery to conflict resolution.

- Modify project details (schedule, budget, personnel, timing) as needed due to changes, risks, issues, and problems.

- Manage and integrate work of the project that is outsourced to other companies or subcontractors.

- Navigate team development from forming through storming and norming to high performing for the benefit of the existing project and future projects.

■ Integrate the final product, service, or intended outcomes of the project.

All of this work occurs whether the project team is co-located or virtual. What is different when it comes to virtual project management, however, is the situational elements, the contextualization of the work, and how project managers and teams engage one another.

Not long ago, Sebastian Bailey, co-founder of Mind Gym, detailed the five killers in virtual work in a *Forbes* article.[12] The five killers include:

1. Lack of nonverbal, face-to-face communication

2. Lack of social interaction among the team

3. Lack of trust

4. Cultural clashes

5. Loss of team spirit

Clearly, each of these five killers of virtual projects has little to do with the blocking and tackling associated with basic project management practices. Rather, these killers all relate to the people side of a virtual team.

Moving from Storming to Norming and Performing

It is during the executing stage of a project that the full team is resourced and engaged in interdependent work. It's also during execution that project managers must lead the team from the forming stage of team development work to high performing.

Chapter 3 detailed the characteristics of a high-performing team. The five distinguishing factors of high-performing teams relative to other teams include:

1. A deeper sense of purpose

2. Relatively more ambitious goals

3. Better work approaches

4. Mutual accountability

5. Complementary skill set (and, at times, interchangeable skills)

Achieving high performance is a challenge for even the most seasoned project managers. Conducting such work is much more challenging for managers overseeing virtual project teams.

Many of us are likely familiar with immersion programs. Whether it is a French immersion for a student from another country to become fluent in French and the French culture or a public safety community immersion in which police, fire, and other safety professionals purposefully immerse themselves in an area of the community to build awareness, establish common goals, build rapport, and establish trust. Immersion programs are similar in that they are strategies to quickly understand the community, align with its culture, and establish common ground and confidence. An immersion program would be great for virtual team development. Actually, nothing would be better. However, costs prohibit such a strategy for project planning and executing.

How can virtual project managers develop individuals and the team when they are not face-to-face and immersed with one another? The first step is to understand that project team member development is the responsibility and the challenge of project managers. Many project managers fail to realize their role in team development. Many assume this is the role of the human resources department or team members' direct supervising managers. Best-practicing project managers do indeed realize their role and, perhaps more important, realize the correlation between team development and project performance—from individuals to the collective team.

Beyond awareness of role responsibility, how does this development get done? First, let's start with how it *doesn't* get done. Team development success is not found in the latest technology. Sure, technology can be an enabler to team development, but often managers seek a technology solution when the problem is not a technology problem. Most problems associated with virtual project team development exist because of one of these reasons: (1) poor interpersonal skills, (2) lack of process discipline and consistency, or (3) team development is managed as an add-on rather than integrated into the team's work.

Based on these problems, the short answer for how best-in-class project managers develop their virtual teams is that they leverage interpersonal skills to know their team members and align developmental opportunities among team members, integrate team development into project work as an expectation and rule, and they spotlight individual knowledge in a way that enables everyone to teach others as much as they learn from others.

Team learning and team development is a social journey. Even if some individuals prefer self-directed learning and activities, they still learn from others in the process. For this reason, virtual project managers must establish a team culture of sharing lessons and best practices as a form of teaching and learning. The old Latin saying is true: "The best way to learn is to teach."[10]

Why is this education so important during the executing phase of a project? For virtual teams, it's because of the need to be intentional and purposeful. Co-located project teams can get together (face-to-face) every week or so to review the knowledge area plans in a status meeting. During that meeting, team members can discuss risks relative to task-level work, milestones, and deliverables and how team members can support one another. They can also (face-to-face) sense frustration or conflict among one another and work through the conflict or schedule a separate meeting to resolve issues. In so doing, the team develops and naturally navigates the stages of team development. Additionally, being co-located allows team members to have coffee breaks together, or meet for lunch, and see what each other values; for example, displaying sports paraphernalia, family vacation photos, or other items in their cubicles allows others to see the personality of the person. All of this helps team members to get to know one another, understand one another, build trust and rapport, which cascades from personal awareness to team member camaraderie. These team member interactions all facilitate team development. Obviously, this rapport building is much more difficult among virtual team members. That's why virtual project managers have the additional role of intentionally seeking ways to build rapport within their virtual project teams.

Whether managers are leading a co-located or virtual team, they want the team to be high performing. It's from this point that collaboration occurs and further enables cross-team integration of work to be completed most efficiently and effectively. Listed in the box below are some additional suggestions from Dale Carnegie for leveraging the power of team members working together.

Lessons We Can Learn from Dale Carnegie

Perhaps the most prominent 20th-century figure to develop content on how to work with one another was Dale Carnegie. When it comes to influencing people and leveraging the power of people working together, he suggested that we become friendlier persons. To do so, he recommended these 10 points:

1. Don't criticize, condemn, or complain.
2. Give honest, sincere appreciation.
3. Arouse in the other person an eager want.
4. Become genuinely interested in other people.
5. Smile.
6. Remember that a person's name is to that person the most important sound in any language.

7. Be a good listener and encourage others to talk about themselves.
8. Talk in terms of the other person's interest.
9. Make the other person feel important—and do so sincerely.
10. The only way to get the best of an argument is to avoid it.

To synthesize and highlight these principles, show that you care, be considerate, and actively listen. These interpersonal skills can mean the difference between mediocrity and success, between failing and winning.[13]

Influencing Virtual Stakeholders

The people side of project execution also includes the work virtual project managers perform associated with influencing a project's stakeholders. A *stakeholder* is commonly understood as anyone who has a vested interest in the outcome of a project. More important, for project managers, a stakeholder is anyone who can influence, either positively or negatively, the outcome of their project. This includes people and groups of people inside and outside the organization. Stakeholder management is a process with which project managers can increase their acumen in managing the political, communication, and conflict resolution aspects of their projects to ensure a positive outcome.[14]

Quite often, the accountability for project success relies more and more on the interpersonal abilities of project managers—those who have limited positional power within the organization yet still own the responsibility for project success. Project manager empowerment in part comes from building strong relationships and successfully influencing key stakeholders.[15]

Stakeholders are many and varied on projects, and they come to the table with a variety of expectations, opinions, perceptions, priorities, fears, and personal agendas that many times are in conflict with one another. One of the facts of virtual projects is that not all stakeholders are physically present and visible to project managers; nevertheless, they can exert significant influence over the project from afar. The challenges in working with stakeholders tend to increase in the virtual environment because of differences in language, time zones, cultural norms, and business practices (see the box titled "Managing Virtual Stakeholders).[16] One virtual project manager we spoke with had an interesting perspective on virtual stakeholders:

Stakeholders = Issues

Virtual stakeholders = Issues lying below the surface

The challenge, therefore, is to find a way to manage this cast of characters *efficiently* in a way that does not become all consuming. Fundamental to efficiency is being able to identify and separate the highly influential stakeholders and then create and execute a stakeholder strategy that strikes a balance between their expectations and project realities.[17]

Managing Virtual Stakeholders

A focus group working with the Association for Project Management reviewed the challenges associated with stakeholder management for virtual project managers and outlined best practices to overcome each challenge. The challenges unique to virtual project managers include time zone differences, language barriers, and logistics. For stakeholders to be managed properly, project managers must have a relationship (ideally with good rapport) with each stakeholder. Relationships are difficult to manage across distances. To mitigate time zone differences, the focus group recommended "vigorous planning" of stakeholder activities within their working hours and planning ample, regular meetings with each stakeholder to build relationship and momentum.

Language barriers can cause problems with meaning and sense making. To mitigate such issues on virtual projects, the focus group recommends project managers assign a site liaison who speaks the local language and acts as a translator. Implementing this recommendation may increase the project budget, but not doing so can be shortsighted relative to downstream project delays caused by stakeholders holding up project decisions because of misinformation or, more likely, misinterpreted information.

Nothing beats face-to-face communication and collaboration. When arranging such meetings, logistics can be a challenge. To mitigate this unique issue, the focus group recommends that virtual project managers research and plan far ahead for the complicated journeys to stakeholder sites and utilize all remote communications methods possible to minimize the need to travel to site. The latter recommendation is especially valuable for virtual project managers unable to travel due to budget constraints.

Delegating Stakeholder-Influencing Duties

Due to the distributed nature of stakeholders on virtual projects, virtual project managers must rely on a number of their team members to share in the project stakeholder-influencing duties. In particular, project managers must tap influential team members who are geographically located near key stakeholders and can establish physical presence with them.

We must recognize, however, that virtual project team members are selected for their expertise in a particular specialty required for the project, regardless of their geographic location and their stakeholder-influencing capabilities.[17]

Selected team members must therefore concentrate on building collaborative relationships and alliances with their local stakeholder community. To do so effectively, the team members must be skilled in relationship building, stakeholder analysis techniques, and negotiating. In addition to distributed team members acquiring the necessary skills to influence project stakeholders, project managers must take care to choose team members who also

possess an innate ability to communicate with people at all levels of the organization. Obviously not all team members will possess this innate ability or be able to acquire the skills necessary to influence others, so project managers must evaluate and choose stakeholder liaisons carefully. Once the liaisons are chosen, project managers also must take care to create an atmosphere in which team members are fully empowered to influence stakeholders on behalf of the project manager.

Instituting Common Methods and Tools

Since effective stakeholder management on a virtual project requires delegation of influencing activities to a number of project team members, it is essential that virtual project managers establish and oversee a common methodology and set of tools. This means there should be commonality in the way team members identify, categorize, and analyze key project stakeholders, regardless of geographic location.

The primary goal of stakeholder management activities is to establish alignment to the strategic goals, intended business benefits, project objectives, and success criteria of a project.[17] Unless there are only a very small number of project stakeholders, however, it is unrealistic to believe that there will not be conflicting opinions and interests between stakeholders. Such conflicts are common, and for this reason, it is important to develop a stakeholder strategy for the project to ensure that the right team members are influencing the right stakeholders. Developing a stakeholder strategy requires identifying all project stakeholders, analyzing the stakeholders' interest in and influence on the project, and then determining how to influence the key stakeholders.

Identifying Stakeholders

Effective stakeholder management begins with identification of all stakeholders associated with a project. It is important that the list of stakeholders be comprehensive to identify all players who may have a vested interest in the outcome of the project.

Regardless of whether the project team is co-located or virtual, when identifying stakeholders, it is important for project managers to cast a wide net, identifying all individuals and groups impacted by the project. The list of stakeholders will certainly include internal people, and also may include individuals and organizations beyond the boundaries of the organization. It is critical to think about the project impact relative to those in your supply chain and across your ecosystem.

Stakeholder identification also includes the categorization of stakeholders into the logical groups to which they belong. Such categories may include senior sponsors, executive decision makers, team members, and resource providers, to name a few. It is important to realize that some stakeholders may belong to multiple groups. The intent of stakeholder categorization is to bring structure to the stakeholder list based on common interests in the project.[18]

Tools such as stakeholder maps are common (see Figure 4.7) and can be effective in helping project managers identify the various stakeholders.

Category	Name	Function	Role
Management	Sue Williams	Product line manager	Sponsor
	Ajit Verjami	Software dept mgr	Resource provider
	Mark Williamson	Customer support mgr	Subject matter expert
	Brian Vigassa	Strategist	Champion / Consultant
Professional Services	Chris Heidler	Procurement	Subject matter expert
	Anna Tamara	Legal	Subject matter expert
	Fern Wilde	Contracting	Subject matter expert
Client	Steven Cross	Director	Client
Vendors	Andy Mulchao	Software vendor	R&D manager
	Fern Wilde	Contracting	Subject matter expert
Subcontractor	Matty Knowles	Test contractor	Manager
	Eric Innis	Test contractor	Contracts manager
	Tuan Ngyen	Test contractor	Test engineer

Figure 4.7: Example Stakeholder Map in Table Format. Source: Russ J. Martinelli and Dragan Z. Milosevic, *The Project Management ToolBox*, 2nd ed. (Hoboken, NJ: John Wiley &Sons, 2016).

The comprehensive stakeholder list should include internal stakeholders, such as top managers, project governance board members, department or functional managers, support personnel (accounting, quality, human resources), the project team, and any external organizations or contractors contributing to the project.

External stakeholders should also be listed and may include contractors, vendors, regulatory bodies, service providers, and others. The objective of stakeholder identification is to include anyone who might have an influence on the outcome of the project.

Analyzing Stakeholders

Project stakeholders can be many, dispersed, and varied in their viewpoints and characteristics. It is important for project managers to do a good job in identifying stakeholders, but stakeholder identification by itself has limited value. Developing a deeper understanding of project stakeholders' interests, opinions, and viewpoints is the necessary next step in the stakeholder management process. This step is commonly referred to as stakeholder analysis.

Stakeholder analysis activities involve determining what type of influence each stakeholder has on the project, such as decision power, control of resources, or possession of critical knowledge, and their level of allegiance to the project. In other words, would the stakeholder prefer the project to succeed, not to succeed, or is he or she indifferent about the outcome of the project?

The purpose of stakeholder analysis is to enable project managers to identify the individuals and groups that must be interacted with in order to accomplish the project goals. Effective stakeholder interaction is supported by thorough stakeholder analysis activities that allow project managers to develop a strategy to:

- Identify strategic interests that the various stakeholders have in the project to negotiate a *common* interest.

- Develop plans and tactics to effectively negotiate competing goals and interests between stakeholders.

- Secure active support from project champions. (See the box titled "What Is a Project Champion?")

- Devise activities to neutralize or prevent the negative actions of nonsupporters.

- Allocate personal and expanded resources to engage with key stakeholders.

What Is a Project Champion?

A project champion is an informal, but important role on a project. Also called the project advocate, a champion is a key stakeholder who is a strong supporter of the project and continuously communicates its benefits and value to other stakeholders. The project champion often serves as a liaison to top management and is effective in communicating at all levels of the organization. He or she also helps project managers navigate the political landscape and serves to remove obstacles that are or may prevent progress.[19]

Due to the champion's position and influence within an organization, he or she normally understands the concerns, goals, viewpoints, and opinions of the other stakeholders. For this reason, the champion is an important individual for project managers to work with during stakeholder analysis activities.

When looking for a project champion, project managers should search for individuals who are:

1. *Respected*. A project champion should be someone who is trusted and whose viewpoints and opinions are sought out and considered.

2. *Influential.* A project champion is someone who can prompt action on the part of other stakeholders without relying on his or her positional power.

3. *Strong communicators.* In order to promote the value of the project and influence action, a project stakeholder has to be a strong communicator.

4. *Politically savvy.* A project champion should be knowledgeable about the formal and informal political landscape and effective at using his or her political capital to realign personal agendas to the goals of the project.

Table 4.2: Example Stakeholder Analysis Table					
Name	**Assumed Role**	**Expectations**	**Reservations**	**Provides to Project**	**Decision Control**
Sue Williams	Sponsor	Project meets all business and execution goals	Firm's ability to develop the new capability	Direction and decisions	Gate approvals
Ajit Verjami	Department manager	No expectations for this project	Believes another project provides a better solution	Resources	Resource allocation decisions
Steven Cross	Client	Project will be completed under budget	Timeline is very aggressive	Funding	Gate approvals
Danielle Carvalho	Subject expert	Project will stay on schedule and complete on time	Already committed to two other projects	Time and expertise	None
...	...	...	...	...	...

Many of the stakeholder analysis activities are about prioritizing project stakeholders. A small subset of stakeholders, commonly called key or primary stakeholders, possess a significant amount of organizational influence to either advance a project or inhibit its progress. Either way, these stakeholders have to be identified through a filtering and prioritization process. A stakeholder analysis table is an effective tool for analyzing the various project stakeholders. Table 4.2 provides an example.[20]

The table becomes the single source of information about the project stakeholders for further stakeholder analysis. To begin the analysis, project managers should use the information within the stakeholder analysis table to focus on four key pieces of information:

1. Determination of the key project stakeholders.

2. Assessment of stakeholder alignment to the project goals, scope of work, and project outcomes.

3. Identification of potential conflicts of opinion between stakeholders.

4. Identification of project advocates and non-supporters.

It is far better to understand if stakeholders agree with the project goals identified and documented early in the project life cycle instead of waiting until it is too late to make adjustments to the goals or to work with stakeholders to get alignment.

Virtual project managers cannot afford to be naive in thinking that all stakeholders want their project to succeed. Unfortunately, this is not always the case. Because of this, project managers can use the stakeholder analysis table to begin determining who the supporters and nonsupporters are. Specifically, close attention should be paid to the information that is contained in the "reservations" portion of the table. Typically, nonsupporters bring their reservations to light during discovery conversations with project managers. (See the box titled "Key Questions to Ask Stakeholders.") Normally, an additional level of analysis is needed to determine if stakeholders who are not advocates may in fact become inhibiters to project success. However, at this stage, project managers should be able to separate advocates from nonsupporters. With information contained in the stakeholder analysis table in hand, project managers and their virtual stakeholder management proxies can begin to craft a stakeholder strategy for the project.

Key Questions to Ask Stakeholders

Understanding the needs, expectations, and potential issues of project stakeholders is crucial to the success of any project. The work project managers do to familiarize themselves with stakeholders and learn as much as possible about them may make or break the project.

The following questions can serve as a starting point for collecting the information to fully understand and analyze stakeholder interests:

1. Who will receive the output of the project?
2. What direct benefit will the recipient gain from the project output?
3. As a stakeholder in our project, who will be working with you from the project team to execute the project?
4. Who are the subject matter experts that should be consulted?
5. Are there other stakeholders who may not favor this project?
6. Who owns the budget responsibility for this project?
7. How does the stakeholder define project success?
8. What problem does the project solve, or what opportunity does it present?
9. How does the stakeholder see his or her role in the project?
10. What worries the stakeholder about the project?
11. Are there any suspected conflicts of interest?
12. What information does the stakeholder need about the project?
13. Is there an alternative project that may be a better investment?

Creating a Stakeholder Strategy

Stakeholder analysis is about sense making. This means understanding the significance of the information gained about the various project stakeholders. Project managers then use the significance of the information to develop a strategy for engaging and managing the right set of stakeholders—those who have power and influence to affect the outcome of the project.

Most of the literature on stakeholder management classifies primary stakeholders as those with both high power and strong allegiance to the project. This is not completely accurate because it leaves out the most potentially dangerous stakeholders—those with high power and negative allegiance to the project. These people also need to be considered primary stakeholders. Figure 4.8 helps to illustrate why. Stakeholders with high power and negative allegiance require significant engagement.[21]

If project managers use the power/allegiance grid to map their stakeholders, a core stakeholder strategy will begin to emerge. The strategy should consist of a communication and action plan for each of the primary stakeholders. It should also keep the project advocates engaged, describe how they can be used to influence others, and plan how to win over or neutralize stakeholders who are not current advocates. The stakeholder strategy should consider:

- Which stakeholders require the most attention?
- Is there a need to influence a change of allegiance for any stakeholders?
- What message needs to be delivered to each stakeholder?
- Can any stakeholders be leveraged as champions?
- What is the best method and frequency of engagement and communication with each stakeholder?
- Does the strategy reflect the interests and concerns of each stakeholder?

The stakeholder strategy likely will identify the need to focus attention first and foremost on some of the most difficult stakeholders. Stakeholder engagement activities will test the courage of project managers who need to be brave and bold when faced with building relationships with stakeholders who are not fans of the project or may be professionally threatened by the project's outcome. Project managers must not follow the human tendency to avoid these stakeholders. Rather, they should seek them out and, most important, listen to what they have to say. Only by listening can project managers begin to find middle ground to use as a means to positively influence stakeholders.

I Forgot to Check the Box

All projects have both obvious stakeholders and nonobvious stakeholders. Because of the lack of physical presence between stakeholders residing at distributed sites, it is likely that more nonobvious stakeholders exist on virtual projects. Stakeholder identification, analysis, and strategy can only pinpoint needed actions for the obvious stakeholders. Surprises associated with nonvisible, nonobvious stakeholders are likely to occur, as they did for Jeremy Bouchard.

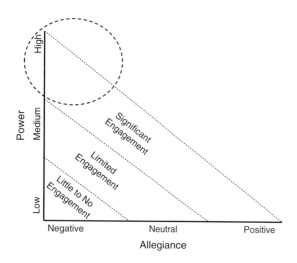

Figure 4.8: Power/Allegiance Grid

"I forgot to check the box that stated the product I was shipping to Japan for system testing was not for resale. Turns out that was a crucial oversight that had significant impact on the Sitka project. Because the product was in final prototype, it *looked* like a completed product that could be sold, even though it wouldn't *perform* like a completed product. The customs agent in Japan who inspected one of the products couldn't tell the difference by looking at it, and he therefore assumed it was not merely an engineering prototype, but could be sold. That raised his suspicion, and our systems test prototypes were quarantined in customs. Three weeks: That's what it took to submit the required paperwork, get it reviewed by Japanese customs, and get a decision to release the test prototypes from quarantine. Three weeks of lost schedule and project productivity. I learned that even a customs agent in a foreign country can be one of my project stakeholders—sight unseen."

Assessing Virtual Project Execution

The virtual project execution assessment is really a readiness assessment that may be conducted several times during project execution. Depending on the duration and complexity of the project, project managers may want to conduct the first assessment 30 to 60 days after the start of executing. The reason for this is to make sure the team gets off to a fast and productive start, to ensure work plans designed in the planning phase are working properly in the executing phase, and perhaps to show early wins.

Additionally, the assessment may be used toward the end of the executing phase of work to ensure proper hand-off of project outcomes from the project team to the operations team. Again, depending on the duration and complexity of the project, project managers may conduct assessments 90 days prior to hand-off and then again at 30 and 60 days after hand-off. The goal of interval assessments such as these is to have the project managers, the team, and key stakeholders see "no" responses in the assessment moving to "yes" responses over time.

It is important to note that many of the assessment items are likely to yield a "no" response in the first 30 or 60 days of execution. That is intentional. For each "no" response, there should be an action plan and owner. It is a risk mitigation tactic to help the project manager ensure the project team has the resources and planning documents needed to achieve project success.

Virtual Project Execution Assessment

Project Name: _____

Date of Assessment: _____

Assessment Completed by: _____

Confidential Assessment: _____ Yes, confidential

_____ No, not confidential

Assessment Item	Yes or No	Notes for All No Responses
Resource Management		
Financial resource needs are known, allocated, and appropriate for the successful execution of the project.		
Personnel resource needs are known, allocated, and appropriate for the successful execution of the project.		

Assessment Item	Yes or No	Notes for All No Responses
Physical resource needs are known, allocated, and appropriate for the successful execution of the project.		
Roles and responsibilities are documented.		
Any impact on personnel resources has been planned with the office of human resources and strategies are agreed on.		
Assumptions		
Base assumptions from the outcome of planning are documented.		
Each project assumption has an assigned validation owner.		
All assumptions have been positively or negatively validated.		
Risk events associated with untrue assumptions have been documented.		
Stakeholders		
All stakeholders (internal and external) have been identified.		
A stakeholder assessment has been completed.		
A stakeholder strategy has been developed and documented.		
All team members with responsibility for influencing stakeholders have been notified and assigned specific stakeholders.		
All team members with stakeholder responsibilities have been trained.		
Governance		
A project governance system has been established.		
Monitoring and reporting expectations have been documented and communicated to the team.		
A common status format and content has been established for distributed team members.		
Distributed team leads are using a standard status indicator.		
A project dashboard or equivalent tool is in use to track overall project status.		
The project objectives are documented and are being monitored on a consistent cadence.		
Project key performance indicators are documented and are being monitored.		

Assessment Item	Yes or No	Notes for All No Responses
Program reviews with senior leaders are being held at a regular cadence.		
Change Management		
All requirements have been detailed in a way that offers a solution that meets the organizational needs for the project and associated change.		
Requirements are under change control.		
A formal change management process is in place and in practice.		
The organization has a culture that embraces change, but validates the need for change.		
Change management protocols have been documented.		
Changes are logged in a common database and change communication channels are established.		
Changes are tracked and reported as part of the project governance system.		
Additional Topics		
Project managers and business personnel have discussed and planned for any integration conflicts with other projects and/or business initiatives.		
Risks are identified and managed with clearly documented mitigation plans.		
Communication processes, channels, and media are known for the project and are functioning effectively.		
The implications of this project and associated changes to the rest of the organization have been communicated to the team.		

Findings, Key Thoughts, and Recommendations

Notes

1. William J. Hamersky, "Business Governance Best Practices of Virtual Project Teams," Walden University, College of Management and Technology, 2015. http://scholarworks.waldenu.edu/dissertations/238/.

2. Roger Dunn, "Development Governance for Software Management," IBM developer Works, October 16, 2007. https://www.ibm.com/developerworks/rational/library/oct07/dunn/.

3. Project Management Institute, "Delivering Value: Focus on Benefits During Project Execution." PMI's *Pulse of the Profession* Report, 2016.

https://www.pmi.org/learning/thought-leadership/pulse/focus-on-benefits-during-project-execution.

4. Russ J. Martinelli, James Waddell, and Tim Rahschulte, *Program Management for Improved Business Results*, 2nd ed. (Hoboken, NJ: John Wiley & Sons, 2014).

5. Russ J. Martinelli and Dragan Z. Milosevic, *Project Management ToolBox*, 2nd ed. (Hoboken, NJ: John Wiley & Sons, 2016).

6. Martinelli and Milosevic, *Project Management ToolBox*.

7. Ibid.

8. Ibid.

9. Ibid.

10. Parviz F. Rad and Ginger Levin, *Achieving Project Management Success Using Virtual Teams* (Plantation, FL: J. Ross, 2003).

11. Martinelli et al., *Program Management for Improved Business Results*.

12. Sebastian Bailey, "How to Beat the Five Killers of Virtual Working," *Forbes*, March 5, 2013.

13. http://www.forbes.com/sites/sebastianbailey/2013/03/05/how-to-overcome-the-five-major-disadvantages-of-virtual-working/#53ba8e3d53d7.

13. http://www.dalecarnegie.com/assets/1/7/GoldenBook_English.swf.

14. Martinelli et al., *Program Management for Improved Business Results*.

15. Martinelli and Milosevic, *Project Management ToolBox*.

16. Rad and Levin, *Achieving Project Management Success*.

17. Martinelli and Milosevic, *Project Management ToolBox*.

18. Ibid.

19. Pernille Eskerod, *Project Stakeholder Management: Fundamentals of Project Management*. (Burlington, VT: Gower, 2013).

20. Martinelli and Milosevic, *Project Management ToolBox*.

21. Rad and Levin, *Achieving Project Management Success*.

5

LEADING THE VIRTUAL PROJECT TEAM

When a firm begins to perform its project work virtually, success is often due to heroic efforts on the part of the project manager and a number of his or her team members. Achieving success *consistently* becomes a shared responsibility between project manager, project team, and senior leadership. It is the job of the senior leaders to establish a virtualization strategy and to remove any barriers to virtual execution success. The company's project managers are responsible for consistently ensuring their virtual projects are executed successfully and that business value is captured. If both pieces of the organization work effectively together, a continual increase in virtual project maturity and repeated success can be expected. As we underscore in this chapter, repeatable success in virtual project execution has to do with solid project management practices plus strong project team leadership.

As discussed earlier, all project managers have two primary roles: managing the project and leading the team. Many project managers experience a shift in the amount of time and effort spent in these two roles when they move from managing traditional projects to managing virtual ones. As Figure 5.1 illustrates, the balance of effort shifts to the right of the management-leadership continuum for a virtual project.

This is not to indicate that team leadership is more important than good project management practices. Quite the contrary, project management practices have to be performed effectively no matter what type of project is being managed. The difference is that for a virtual project, work has to

be performed through team members who are much less accessible and are loosely connected organizationally. Therefore, influencing from a distance is required, communication from a distance is required, and monitoring of work from a distance is required.

In short, the distance factor caused by the team's distribution requires additional effort and strong team leadership practices on the part of project managers. Contrary to what some project managers have come to fear, however, managers of virtual projects do not need to acquire a completely new suite of skills and competencies to succeed in leading virtual project teams. The foundational elements of effective team leadership apply whether project managers are leading traditional project teams or virtual project teams. As Jeremy Bouchard and others have come to learn, success begins with proficiency in the basic principles of team leadership and then an understanding of how to extend the leadership principles to apply to virtual teams. Over time, the basic leadership principles are augmented with additional processes, tools, and skills to increase effectiveness, given the cultural, distance, time, and communication challenges that virtual projects present.

Understanding the basics of project team leadership begins with understanding the various roles project managers play when leading teams of project specialists. It all begins with setting a common direction to channel team energy and work output.

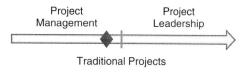 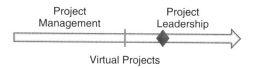

Figure 5.1: Virtual Projects Require a Shift in Focus

Be the Project Compass

Project managers, acting as project leaders, are responsible for setting the direction for the project. Setting a common direction for a virtual team is a vital role, as virtual team members often scatter their work effort in many different directions. Like all project teams, virtual teams contain a high level of human energy that drives team members to want to contribute in some way to project success. Without clear direction, that energy will be scattered and un-focused, diminishing the power that can be gained from collective human energy. A simple analogy is that of a search party whose job it is to locate a missing item in a large geographical area (such as a five-acre area of trees and brush). If a search party of 20 people must search on their own to find the item, they will scatter in 20 different directions, and the result will be duplication of effort, territory that will remain unchecked, and a high level of inefficiency of human effort. We have learned that the effort of these 20 searchers is much more productive and successful when they are organized, structured into a line, each given an area of responsibility, and guided by a common direction of search. The search leader in effect serves as the human compass for the search team. In similar fashion, project leaders have to serve as human compasses for virtual project teams. To do so, they must know exactly where they are leading the team and what direction the team needs to be heading. This is accomplished through the creation of a project vision.

Project Vision

In Chapter 3, we discussed the need for a project team to have a common purpose or mission in order

to build team chemistry and help team members attach to a common and meaningful purpose for their specific work on the project team. At times, the project purpose or mission is confused with the project vision. Two key words help to distinguish between mission and vision, *why* and *what*. The project mission defines *why* the project is needed. For example, the mission of a project may be to in-crease market share or revenue, to reduce operating costs, or to enter a new business segment.

The project vision, in contrast, defines the project end-state, or *what* is to be accomplished by the project outcome. The vision, in effect, helps to define the organizational goals to be achieved through the investment in the project. (See Figure 5.2.)

However, while leading project teams, project managers can use the project vision for more than just establishing project goals. One of the essentials of strong leadership is that leaders *pull* on the energy and talents of team members rather than pushing, ordering, or manipulating team members. Leadership is not about imposing the will of the leader, but rather about creating a compelling vision that people are willing to support and are willing to

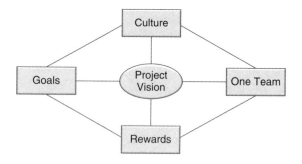

Figure 5.2: Project Vision

exert whatever energy is needed to realize it.[1] The project vision is the vehicle that project managers use to pull the team together and establish a sense of *we* instead of *me*. This "One Team" mentality is important for virtual projects, as it is necessary for all team members to put team goals ahead of functional or organizational goals. Without a compelling project vision, team goals will become subservient to individual and organizational goals. The means by which team members are rewarded has to be designed in such a way that reinforcement of team goals has priority over functional goals.

As the virtual project team begins to view themselves as "One Team" with a common purpose and vision, a unique project culture begins to form. Team culture, a micro culture of both organizational and national culture, is important to establish because it helps to create a unique team identity and operating principles. Team culture is powerful because it is formed by shared experiences that solidify beliefs, values, assumptions, and behaviors. At this point, the team is moving beyond norming and storming to (high) performing. To establish a positive project culture, project managers must first recognize their role in doing so. Second, project managers must intentionally design the means to positively affect the culture of the team by articulating a clear mission, developing a shared vision of the team's identity, and providing team-building opportunities.

Be the Team Conductor

The analogy of the role of project managers as being similar to conductors of orchestras who must keep all players synchronized in time and playing complementary parts from the same sheet of music has been used many times. Just as an orchestra is composed of multiple instrumental elements (e.g., the string, brass, and percussion sections), a project is composed of multiple functional elements (e.g., design, development, and manufacturing). Alignment of all project elements is a critical factor in project success and a primary leadership challenge for project managers.

Gone is the time when project team leadership is authority driven with work being performed through a series of directives. Today's era of project team leadership is more diffused, collective, and team-based, where directives and decisions are participatory, collective, and democratic. Project managers success in the role of team leader is dependent on how well they facilitate the alignment of interests, work activities, and the collective outcome of the project. (See Chapter 9.)

Product and service solutions are now too complex to be developed by a single expert. The integration of work from a team of specialists who focus on creating their piece of the solution is required. However, specialists do not by nature desire to work in a collaborative and participative manner. They would prefer to be left alone to create their piece of the solution and hand it off to someone else to integrate and use.

Unfortunately for specialists, this type of development has been proven to be highly inefficient and ineffective and has led to the need for collaborative cross-functional work teams. Surrounded by specialists, the role of project team leaders becomes one of initiating and driving continuous, cross-team collaboration.

When leading a highly distributed team, project managers have to work even harder to drive effective participation and collaboration between members who are separated by time and distance. This is due primarily to the fact that participation and collaboration—as well as integration of work output—has to be performed asynchronously and electronically.

The physical distance that separates virtual team members limits the amount of synchronous collaboration. Because face-to-face meetings and discussions are limited, time delays occur due to the additional iterations of communication and work that result from trying to collaborate electronically and at different times. Relatively routine tasks, such as scheduling a meeting, can become complex when one person's workday is beginning while another is sitting down to dinner.[2]

Time differences and physical separation can also allow some of the more introverted members of a team to get lost in action—meaning their team participation and collaboration can become lost due to their innate personalities. Project team leaders must exert focused effort to keep these people active and engaged.

Integration of work output also becomes more complicated on virtual projects. Work is accomplished in a very fragmented manner on a distributed team, and it will remain fragmented unless project team leaders purposefully establish and manage integration of work between members.

Effectively driving collaboration, participation, and integration of work output, therefore, requires a change in both behavior and some processes. To first establish broad participation of team members, team leaders must focus on some critical behaviors. They must be willing to listen first, not tell their team what they think first and then ask for opinions. A more effective way to increase member participation is for team leaders to ask for input and opinions first, facilitate a discussion as a team, and then share their own opinions. By doing so, team leaders are motivating the team through inclusion. Through deliberate provocation of opinion from the team, team leaders help them become more comfortable with participating and sharing their opinions over time.

Another effective approach for driving a high level of collaboration between members of a virtual team is to organize the team's work in such a way that the team members are mutually dependent and they recognize the benefit of that mutual dependency. One of the intangible benefits of creating a project map that clearly highlights team interdependencies as covered in Chapter 2 is that the map demonstrates that a high degree of collaboration *is required* for the team to succeed in creating the project output. Additionally, the mapping process itself helps to build the virtual team as a cohesive entity.[3] This is especially true if the mapping process is performed in a face-to-face session where at least the core members of the geographically distributed virtual team meet, work together, and begin to build professional relationships with one another. Using the project map throughout project execution is an effective way to stress the need for continual collaboration and to orchestrate the integration of work outcomes.

Be the Champion

It is true that a project team needs good direction and a vision of success. It is also true that team members need help coordinating and integrating their work outcomes, especially when they are geographically and organizationally distributed. And, equally important, teams need leaders who will listen to their ideas, concerns, and needs and who will help them take responsibility to meet their own goals.

Good project team leadership, then, is not only about accomplishing team goals and purposes, but also about creating a motivating team environment built on trust and then empowering team members so they can prosper and grow in the process. Project managers truly champion the efforts of the project team members. Team leaders must strive to enable the team to succeed and to set expectations that the team members will make decisions, solve problems, and collaborate effectively. A major factor in the enablement of a virtual project team to achieve self-sufficiency is how well project managers motivate their teams.

Motivating the Virtual Team

It is no surprise to virtual project managers that recent findings show project team member motivation is critically important to success, or that the importance of motivation is equal for both traditional and virtual projects, or that fostering motivation is particularly more challenging on virtual projects than on traditional ones.[4]

Two of the subtleties that have emerged specific to motivating a virtual team are interesting for virtual project managers to consider. First, social interaction has emerged as a factor in sustaining motivation on project teams. Obviously, virtual projects hamper social interaction by nature, and severely so in highly distributed virtual projects. Second, it is recognized that senior-level people on project teams are more difficult to motivate and keep motivated. Having spent a good portion of their career on traditional projects, senior-level employees seem to have a more difficult time adapting to the virtual project environment than less experienced team members. This difficulty can, in fact, become a demotivator. Another factor affecting motivation of senior-level employees is previous fulfillment of motivating factors. As we explain later in the chapter, motivators tend to be intrinsic, or internal, in nature. And these intrinsic motivators do not change much over time. The result is that the longer a person's career, the more likely that his or her internal motivators have already been fulfilled.

With these two subtleties in mind, we explore the subject of motivation on virtual teams from a wider landscape.

The title of this section is quite misleading. It leads readers to believe project managers can indeed directly motivate the members of the virtual project team. Despite the millions of dollars spent each year on motivational speakers and workshops that teach us how to motivate our teams, it is understood that people, including the people on project teams, are in fact *self-motivated*.

Consider individuals on a project team who seem completely unmotivated. What happens to them when they go home to their families, their hobbies, or other important aspects of their lives at the end of the workday? Odds are high that they become *very* motivated individuals. The person does not change, but the climate and atmosphere of their environment in fact changes. What are we trying to convey? Simply put, virtual project team leaders cannot motivate their teams directly; motivation has to come from within the individuals. It is key to understand that the role of team leaders is to create a motivating *climate* for project team members.

Among project team members who are self-motivated, intrinsic factors (those that are personally rewarding) are more important than extrinsic factors (those that provide reward or punishment avoidance) for creating and sustaining motivation. The panel of virtual project team members (a virtual team in itself) who participated in the research for this book confirmed this assertion. When asked what motivates them most in the workplace and project environment, they provided these answers:

- New challenges, being able to try new things and either succeed or fail.

- Seeing their team complete a project successfully.

- Knowing that their part of the project made a big contribution to the overall project.

- Pushing themselves to new levels and learning new technologies and techniques.

- Having the opportunity to work on diverse types of projects.

- Being able to work with and get to know people from around the world.

All of these factors are related to the internal gears that drive each individual with little or no connection to external factors (such as paychecks, incentives, or bonuses). Research shows that intrinsic motivators are even more important on virtual projects. The more virtually distributed a team is, the weaker the social bonds and personal interaction; therefore, the more team members have to rely on internal factors to remain motivated.[5]

This does not mean, however, that extrinsic motivators such as a competitive salary or bonuses for a job well done are not important to keep the individuals working at the same level of motivation. If extrinsic factors are not also considered, they can indeed become *demotivators*. Consider this question: Would you as a project team member remain highly motivated if any or all of these extrinsic factors were present in your project environment?

- The project lacks clear goals and success factors.

- You are asked multiple times to rework your part of the project because the overall project direction continues to change.

- You feel isolated from the rest of the team.
- The work you do lacks challenge.
- You fear retribution if you make a mistake.
- You lack the power to make necessary decisions needed to keep your work progressing.

Therefore, the key point for virtual team leaders to remember is that they are responsible for creating a motivating climate; virtual team members are personally responsible for finding the things that motivate them the most within the environment and following through to project completion.

Creating a Motivational Climate

For project teams to succeed, the climate in which team members work must be positive, inspiring, forward-moving, and motivational. There are many opinions on what *project climate* means, but to us it is what employees feel and experience when they work as part of the project team and become acquainted with the team culture, the project vision, how team members interact with one another, and the project manager's leadership style. If team members like what they feel and experience, they will begin integrating their individual intrinsic motivators with the project. Project managers must create opportunities for achievement, provide adequate resources to succeed, assist the team in knocking down barriers to progress, give recognition for good performance, and help team members feel like they are an integral part of the team.[6]

Differentiating *Climate* from *Culture*

The terms *organizational* or *team culture* and *climate* are often used synonymously. However, although culture is the shared learning among a group that solidifies beliefs, values, norms, assumptions, and expected behavior, climate is more precisely about the perceptions and attitudes employees or team members have about their organization or team.

Additionally to the team culture, climate and organizational considerations, one of the most important factors in establishing a motivational project climate is empowerment. This means that project team leaders are willing to share power with the team to set goals, make decisions, and solve problems. *No one* likes to be micromanaged, and in today's work environment, nothing will demotivate team members faster than a micromanaging team leader and lack of empowerment. Instead of micromanaging, project team leaders must be willing to support team members' actions and efforts, provide recognition when the team succeeds, and coach them through their mistakes.

Remember, however, that empowerment and acting empowered take time as they go hand-in-hand with the process of establishing trust and navigating through the stages of team development. Project team leaders must become confident that team members will act in the team's best interest when making decisions, setting goals, and solving problems. Also, it is unrealistic to expect team members to assume a great deal of responsibility in the early stages of the project if they have traditionally worked in a more command-and-control environment. The transfer of power and responsibility most likely will need to be gradual. In fact, if empowerment is thrust on team members all at once, the result may likely be demotivation.[7] Again, this is due to team members' feelings that they will be chastised if they make mistakes.

A positive climate for motivation is achieved when a fear of making mistakes is nonexistent and project team managers are willing to coach team members through the mistakes. For a virtual project, there is no option but to share power and delegate decisions, problems, and goals in a decentralized manner where they can be acted on quickly.

Assessing Team Motivation

A powerful question is often at top of mind for virtual project team leaders: How do I know if my team is motivated? The normal approach for assessing team motivation is to evaluate the progress toward task completion. The base assumption underlying this assessment approach is that if team members are doing their work, completing it on time and as tasked, it is an indication that they are motivated.[8] It should be pointed out that there has been no proven causality between motivation and a team member's ability or willingness to complete work. The relationship between motivation and work progress simply remains an assumption.

Even though the validity of the work progress approach to assessing team motivation is debatable, the underlying problem that this assessment approach is trying to solve is that because there is a lack of face-to-face interactions on virtual projects, people need to rely on other cues to determine the level of team motivation. Jeremy Bouchard shared the following lesson learned from his first virtual team experience: "Because I was not in a room with my team personally, I had to rely on their perceived enthusiasm that was projected through their voices. I was listening for the degree of responsiveness they demonstrated during our conversations because I couldn't see their body language."

For some ambitious (and we would call them brave) virtual project managers, we suggest a more direct approach—have team members take a motivation climate survey. (See Table 5.1.) Since team members are, by and large, motivated by *intrinsic* motivators, it makes little sense to assess if they are truly motivated. Rather, what needs to be assessed is

Table 5.1: Motivational Climate Survey	Never		Sometimes		Always
I feel empowered to make the necessary decisions within my scope of work to keep the work progressing as scheduled.	1	2	3	4	5
I am given the authority and responsibility to solve problems within my scope of work.	1	2	3	4	5
I feel empowered to assist my teammates in solving problems outside of my direct scope of work.	1	2	3	4	5
I feel empowered to set my own work goals.	1	2	3	4	5
The work I am doing is challenging.	1	2	3	4	5
I feel recognized for my achievements.	1	2	3	4	5
There is an atmosphere of trust on the project.	1	2	3	4	5
The project goals are clear.	1	2	3	4	5
I feel accepted as a valued team member.	1	2	3	4	5
I understand how my work contributes to the overall success of the project.	1	2	3	4	5
I feel encouraged to innovate and create new ideas and approaches.	1	2	3	4	5
I feel I won't be punished for taking risks if I fail.	1	2	3	4	5
COLUMN TOTALS AVERAGE OF ALL COLUMN TOTAL RESULTS					

whether virtual project managers have established a motivational project climate.

Once team members complete the survey, the results can be averaged for each question and then totaled. Obviously, the higher the sum total of the survey, the more motivational the team climate. Besides providing a quantitative assessment of the motivational climate on a team, the assessment can be used to make corrections in areas where the team members feel the motivational climate is low (columns with the lowest totals). For example, team members may not feel fully empowered to set their own work goals. Going forward, the project team leader can work more collaboratively with the team members on setting work goals and ensuring that they stay aligned with the overall project goals and outcomes.

Maintaining Trust

As we explored in Chapter 3, achieving a high level of team performance with a virtual team requires a solid foundation built on trust. Trust is created by the demonstration of proven interaction and consistency of behavior.[9] An unfortunate reality is that trust between the project team members takes time and effort to establish, but can be destroyed quickly and easily. The same is true for trust between project managers and their teams. Maintaining trust on the team, therefore, becomes a critical factor in leading a virtual project team.

In many ways, virtual project managers are the glue that holds the team together. They must set the expectation that all team members demonstrate they are worthy of mutual trust. The following are ways in which that can happen.[10]

Perform competently. With the absence of personal relationships on a virtual team (especially at the beginning of the project), team members will be evaluating each other and the project manager based on how competent they believe a person is within their role on the team.

Act with integrity. Virtual project team members will closely watch and listen to determine whether other team members act in a manner that is consistent with what they say and within their stated values.

Follow through on commitments. Similar to integrity, this means doing what you say you're going to do, whether it is completing a project deliverable, sending an email when promised, or scheduling a meeting in a timely manner. Those who follow through consistently are perceived as being more dependable and therefore earn team members' trust and confidence.

Display concern for the well-being of others. People trust those who are perceived as responsive to the needs of others on the team and within the organization. Team members (and team leaders) who assess how their behaviors affect other team members most likely will be perceived as having more concern for others. In contrast, those who appear to be less sensitive to others will be viewed as less trustworthy.

Behave consistently. Members of a virtually dispersed team look for a higher level of consistency of behavior among their fellow team members as a way to drive out some of the ambiguity within a virtual team environment. Trust on a virtual team is based on behavioral consistency rather than on social bonds. As project leaders have difficulty assessing trust through social indicators, they have to rely on consistency of behavior. The same goes for team members when building trust in the project manager.

One of the best ways to maintain trust is for virtual project managers to set the example. Team leaders who want team members to exhibit specific behaviors must ensure that they demonstrate, through their own actions, those behaviors. For example, project managers who encourage the team to act with integrity, show respect to others, and meet commitments must model those behaviors themselves.

It is difficult to separate trust between members of a virtual project team and team chemistry, as they are tightly interwoven. Project team leaders must try to ensure that positive team chemistry is maintained. Doing this requires a focus on maintaining social presence even though team members may be widely distributed. James Lehey, an experienced leader of virtual project teams, explains that this is not a difficult task.

> Maintaining good team rapport does not require special effort. It only requires consistent effort. I find it beneficial to add social communication time into our weekly team meeting agenda, usually at the beginning of the meeting. The team also has a section on our team site where we share social experience and events. I use these postings as a way to end some of our meetings by asking a person to verbally share their experience with the team.

The sharing of personal experiences increases the social presence on a virtual team, which strengthens personal relationships among team members, which in turn strengthens trust.[11] Team leaders must also be willing to meet face-to-face with team members at their various geographical locations when possible for local team meetings, one-on-one discussions, review of issues, and discussion of project progress. These efforts provide additional opportunities for project managers to reinforce teamwork and trust.

Virtual project managers must treat trust as their greatest asset and realize that it is important to consistently model the behaviors that exemplify competence, connection, and character in leading the team. The best virtual project managers we have encountered realize the need to go beyond establishing expectations that the team demonstrate the trust-building behaviors. The managers also model the behaviors on a daily basis by always standing behind and supporting their teams, never demonstrating favoritism, and accepting full accountability for the actions and results of their teams.

Demonstrating Personal Integrity

One of the surest ways to destroy trust and motivation on any team, either co-located or virtual, is for project managers to demonstrate a lack of integrity.

Team leaders who operate with integrity behave ethically and honestly, have an unwavering commitment to their values and are willing to defend them, are authentic in that they display their thoughts and beliefs through their actions, and consistently demonstrate responsibility and accountability. For proof of the importance of leaders operating with integrity, just review many stories in the press today that demonstrate the effects of lapses in integrity.

For virtual project managers, integrity is rooted in two foundational elements: values and vision. Values are what people stand for, and the team needs to see demonstrable proof of project managers' values in action. Vision is a clear end state of where project managers are taking the team. For project managers, this involves establishing transparency in the project and business success criteria that tell the team what it takes to be successful in their mission. Anyone can call themselves a leader if they have the knowledge and skills to perform the role. But, if their team members and constituents do not believe that that person uses leadership knowledge and skills with integrity, he or she is not really perceived as a leader.

Managing Conflict

It is impossible to maintain a conflict-free team environment, as conflict is a way of life when people work together. Conflicts typically emerge because people do not see everything the same way. Each member on a team will demonstrate personal preferences, personality traits, ideas, and opinions that occasionally lead to conflict among team members. In this manner, conflict is good! When managed properly, the differences in people's points of view can be a source of power for teams because they represent a broader perspective and more possibilities for creative solutions. The positive effects of constructive team conflict include identification of alternatives not yet considered, better solutions to problems, better decisions, more attainable goals, and broader buy-in across the team than if differences are not discussed. (See the box titled "Managing Virtual Conflict."). A basic rule of conflict should be remembered: It is okay to disagree, but it is not okay to be disagreeable.

Managing Virtual Conflict

Shlomit Shteyer, director of Software Delivery at a major networking company, previously served as an experienced virtual project manager and explained to us how she has addressed management of conflict in her teams in the past.

"It's a common practice for software companies to engage with virtual teams. I've worked with multiple virtual teams around the world that combine remotely to serve as one team in trying to deliver a software product. When it comes to keeping the teams working in harmony, there are a few steps I generally take at the outset. For starters, the key to optimizing any organizational complexity begins with some cultural context. Some important questions to then ask yourself are: In which countries are the teams located? How are they used to working? How do they allocate their roles? What is their decision path today? How is their work culture organized?

The reason for understanding the answers to the questions above, are that, as a project manager, most of your challenges will revolve around how to enable and facilitate conflict resolution when you're dealing with:

- Different cultures and locations.
- Communication channels that do not include face-to-face methods.
- Different work and cultural expectations.

Based on my experience, when I sometimes worked with teams in three different locations that needed to build a product together, I found it helped to act as a point person who could mediate between the teams. Taking this role made it easier for everyone to facilitate the sorts of discussions that built trust and provided a framework for how to resolve any inevitable conflicts that popped up. Each time I entered one of these discussions, I took the three viewpoints listed above into consideration. It was vital to understand how each team operated, behaved, and was accustomed to working. Once I understood these factors, it allowed me to facilitate discussions without judgment, while guiding the teams toward healthy resolution."

Of course, team conflict often has negative effects as well. The inability to effectively manage through conflict between project team members is a major source of team dysfunction. Causes of negative team conflict vary greatly depending on the situation and individuals involved. However, two of the most common causes are competing interests among team members and communication breakdowns. When these situations occur, negative team conflicts can result in reduced productivity, missed delivery dates, hoarding or not sharing vital project information, finger-pointing and scapegoating, reduced collaboration, low team morale, and high team member turnover.

Managing conflict on a virtual project can be a more difficult challenge for team leaders. On a virtual team, conflict can occur quickly due to the high potential for miscommunication, can go unnoticed for a longer period of time because most communication is performed electronically and asynchronously, and can be harder to correct due to a lack of direct face-to-face interaction. As one virtual project manager told us, "Problems resulting from a simple language miscommunication can take on a life of their own. A simple email exchange can end up frazzling nerves because of a simple cultural misunderstanding."

Virtual project managers must be hypervigilant in identifying conflict among team members, because conflict cannot be resolved if it is not identified. Most times, virtual project managers are in the best position to notice conflict in email exchanges between team members or detect negative conflict during team discussions and meetings. As one team member observed, "A lot of leaders ignore conflicts between team members, hoping that they will just go away. But, they seldom do; they usually just get worse." Conflicts among team members must be resolved quickly and to the team's satisfaction. Issues and conflicts cannot be left to fester. Avoiding issues with the hope that they will resolve themselves and disappear may backfire on team leaders later in the project.

It is not enough for virtual project managers to merely be good at spotting conflict on the team. They must also be quick to respond so it is less likely to spiral out of control. Being quick to respond, however, does not mean jumping in whenever a debate of viewpoints is occurring on a team. Virtual project managers must learn to get involved within certain boundary conditions, such as when conflict:

- Affects the performance of other team members.
- Jeopardizes achievement of team goals.
- Interferes with team communication.
- Overflows to external stakeholders or partners.
- Involves a repetitive pattern.

Outside of these boundary conditions, team leaders should allow debate among team members to continue if the debate is moving toward a positive outcome. Doing so will help build team cooperation, collaboration, and cohesiveness.

Understanding Sources of Conflict

It is also important for virtual project managers to become skilled in determining conflict type once a conflict is identified. Conflict type will determine the general course of action that team leaders will want to follow.[12] Listed below are the key sources of conflict that can occur on virtual project teams.

Task conflict. Task conflict results from differences in viewpoints and opinions pertaining to what the team is tasked to do. This type of conflict can be beneficial, and team leaders should encourage varying viewpoints. Task conflict can improve decision quality and ensure that the team is working on the most important set of tasks. Virtual project managers must act as facilitators by embracing debate, but also steer the outcome toward achievement of the intended business results.

Process conflict. Process conflict involves debates regarding how the tasks are performed on the team or how resources are delegated to the tasks. In general, process conflict can be beneficial in improving the effectiveness and efficiency of the team's work, but continued debate can easily turn process conflict into relationship conflict. The virtual project manager must be vigilant in hearing all viewpoints concerning a process, but then be decisive in determining the course of action and set expectations that the process will be followed.

Facts and data conflict. Fact or data conflict occurs when team members have different perceptions of the facts. Often team members use assumptions to jump to conclusions and become emotionally involved instead of asking for more facts or data. Virtual project managers can help resolve this conflict by asking for additional information and keeping the discussion focused on the data without the emotion.

Priorities conflict. Priority conflict stems from team members not being clear on project priorities. If priorities are not clear, team members will debate what tasks are most important to work on, and frustration quickly will escalate due to the interdependent nature of work outcomes on a project. Virtual project managers must be diligent in keeping team priorities clear and in consistently communicating the priorities, especially any changes in them, to the team.

Relationship conflict. Relationship conflict is an awareness of interpersonal differences. This may include personal differences or hostility and annoyance among team members. Since relationship conflict has a negative effect on individual and team performance, virtual project managers must be quick to identify the root cause of the conflict and take measures to begin resolving it. Also, as team members develop a greater awareness of each other through virtual social interaction, relationship conflicts can be minimized and cooperation can more likely be attained.

Values conflict. Conflicts concerning values are the most difficult to resolve because values are deeply personal. Team members grow up with and hold onto different values based on their backgrounds as well as where they live. This is especially evident when the virtual team is internationally distributed. Virtual project managers cannot expect that a group of people from different national cultures and backgrounds will share the same values. Managers should continually work to help team members stay focused on what they can agree on rather than letting them become polarized.

With all types of conflict, virtual project managers must allow individuals the opportunity to express their differences and allow team members to establish resolution as the first approach. It is not a matter of *who* has the best ideas or solutions that is important; What is important is finding the solution that will provide the greatest benefit to the team. Doing this requires the establishment of an open team environment and an understanding of the team guidelines for conflict resolution. It may also require coaching on the part of project managers. If the individuals in conflict are not successful in finding a solution among themselves, team leaders will need to intervene (see the box titled "Tips for Resolving Team Conflict").

Conflict management presents one more argument for taking the opportunity to establish face-to-face contact between members of a virtual team. Evidence shows that on geographically distributed teams, interpersonal bonds are lower, team cohesiveness is lower, members are less satisfied with cross-team interaction, and in general people like each other less than compared to traditional teams.[13] These are all factors created by a lack of trust among members of the team and lead to a higher incidence of negative team conflict. In order to trust one another, people must know one another. The surest way to help the geographically distributed team get to know one another is to get them together physically and allow personal relationships to form.

Tips for Resolving Team Conflict

- *Listen.* Only by listening can someone try to understand the other team member's point of view.
- *Separate people from problems.* When disagreements occur, it is common for those involved to take things personally. By maintaining focus on the problem—the source of conflict—the disagreements will become less emotionally charged.

- *Be accepting.* Accept the right of other people to have a different opinion. Acknowledging their right does not mean you have to agree; it's about demonstrating respect.

- *Focus on the present.* Don't bring up past mistakes or results as a way to discredit others' ideas. Although it is always good to learn from the past, it should not be used as a basis of argument.

- *Be assertive but not aggressive.* Be direct and succinct in communicating your opinions in order to state your viewpoint. Don't be aggressive in trying to *force* your viewpoint on others.

- *Be open minded.* Have an open mind, at least initially, to allow for the opportunity for others to build on your ideas.

- *Don't compete.* People like conflict because it speaks to our competitive nature. Competition, however, creates a win/lose situation. Alternatively, compromise and collaboration with team members create win/win situations.

- *Walk away.* If emotion hits a high point or an impasse is reached, take a break and come back later with a fresh perspective.

Recognizing the Virtual Team

Providing recognition and rewards on a project can be risky business, even more so for a geographically distributed team. The intent should always be to increase the visibility of the good work being accomplished by the team through recognition.[14] However, care has to be taken not to create a reverse effect where team members become demotivated, and trust is compromised.

Co-located teams have traditionally used a variety of approaches to celebrate and recognize project team success ranging from pizza parties, to cash, to movie tickets, to formal organizational awards. However, with virtual project teams, identifying effective and appropriate means to recognize team and individual contributions and successes meaningfully is more difficult. To complicate things further, some recognition approaches that are appropriate in one cultural setting may not be appropriate in another.

Project managers should ensure that team accomplishments are consistently recognized via communication to the sponsor and other key stakeholders. Managers can extend the recognition to the broader organization by tapping into the organization's communications or public relations groups. Project managers should look to accomplish increased public awareness of the team's work and to create a sense of team identity among team members.

Although recognition for virtual teams is harder to administer, many believe it is more important than for co-located teams. It is important to ensure that all involved sites are included in any recognition activities. This is very helpful in overcoming some team members' feelings of isolation. Virtual team members need opportunities to celebrate large and small accomplishments to sustain a high level of team performance.

Rewards are a bit trickier than recognition because they usually are individually based and many times are deemed more valuable because they are tangible. Additionally, rewards are more difficult to provide since team leaders usually cannot offer team members pay raises, bonuses, or vacations.

This requires virtual project managers to be more cautious and creative in providing rewards. See the box titled "Recognizing Virtual Team Members" for some suggestions for recognizing virtual project team members for a job well done.

Recognizing Virtual Team Members

Rewarding or recognizing project team members for a job well done helps to build team morale. The project manager's sincere appreciation can go a long way toward increasing allegiance to the team. Gift certificates, team lunches or dinners, a platter of doughnuts, and team events are all common methods for recognizing the efforts of project team members.

But, what if the team is geographically dispersed? Rewarding and recognizing virtual project team members is more difficult and requires more thought. To help stir the thought process, our panel of experienced virtual project managers offer the following suggestions:

- *Informal kudos*. Kick off team meetings with informal recognitions or kudos for team members. It is best if the recognitions come from fellow team members and are presented as a way to pat each other on the back. Never wait to the end of a meeting, as time constraints can prevent getting to this portion of a team meeting. Removing recognitions from the meeting agenda sends the wrong message to the team.

- *Team awards*. Create team awards, such as a "Going the Extra Mile" award or an "Over and Above" award. If the team is multicultural, be aware of sensitivities to individual recognitions.

- *Pick up the phone*. Take the time to call team members on the phone to recognize someone. In a world of email, instant messages, and text messages, taking the time to call and have a conversation with someone is a special way to express your appreciation to a virtual team member.

- *Get your executives involved*. An email or phone call of recognition from a company executive or senior manager puts a golden touch on a recognition. Vice Chairman of NetApp, Tom Mendoza, asks project managers to notify him when someone is deserving of a recognition. He calls 10 to 20 employees each week to personally thank them.

- *Give the gift of time*. Project life can be hectic and time consuming, and personal time can become a rare commodity for team members. Rewarding virtual team members with the gift of time by providing additional days off is an impactful and powerful way to recognize them.

- *Gifts*. Giving gifts can be a bit of a gamble on a virtual project, especially if the team is multicultural. One way to give a gift that cuts through the sensitivities is to send a gift basket to the location where the team member works, and do so in the team member's name. This approach recognizes the team member, but others at the location can share in the reward.

- *Face-to-face meeting opportunities*. Even though there are many advantages and conveniences associated with being part of a virtual team, humans are social beings who value social exchanges. Rewarding a team for a major project accomplishment by providing an opportunity for a face-to-face meeting is a strong way to show appreciation and build team relationships at the same time. This form of recognition will require prior planning and probably a budget.

- *Decorate*. Having an office decorated with balloons, streamers, or banners is a simple, but obvious, method for recognizing either individual or team accomplishments. Of course, this method may be challenging if the individual works out of his or her home office.

- *Don't forget to say thanks*. Many times, simply remembering to say thank you for a job well done is a sufficient form of recognition. The important thing is to express thanks in a timely manner; letting too much time pass between an accomplishment and a recognition diminishes the impact.

To provide recognition and rewards effectively, virtual project managers must take stock in the things that they have direct control over and employ them consciously, cautiously, and consistently. They must always look to recognize and reward team accomplishments over individual accomplishments. More on this in Chapter 9.

Leverage Your Leadership Style

All project managers develop their own leadership style based on what they believe works best for them to obtain successful results on projects they manage. It is commonly accepted that most project managers adapt their leadership style based on their project environment. Team leaders who have managed in the traditional team environment and later transitioned to leading teams virtually have had to adjust their leadership style. Accordingly, doing this is necessary to accommodate the lack of or limited face-to-face time for their team members and the need to use electronic means for nearly all team communication and collaboration. Additionally, their leadership style must accommodate teams composed of representatives with other first languages, cultures, and customs.

Due to the fact that more time needs to be devoted to team leadership in the virtual environment as compared to the traditional project team environment, virtual project managers can benefit from a leadership style that emphasizes the understanding and application of behavioral, social, and emotional skills. These skills aid leaders in sharpening awareness, sensitivity, and understanding of their team members. Various texts discuss a number of leadership styles. Some are more commonly used than others. No doubt, leadership styles will continue to evolve to fit the future virtual team environment. The next sections help to determine the type of leadership style needed for different types of project teams.

Transactional Leadership

James MacGregor Burns was one of the first authors to characterize the transactional leadership style.

According to the transactional leadership approach, the focus of the team leader is on task accomplishment and achieving project objectives. The leader exhibits behavior that is focused on establishing well-defined roles and responsibilities, team member accountability, consistent follow-up on assignments, and monitoring achievement of project milestones. Also, prompt feedback from team members on problems and other issues is a high priority.[15]

It is relatively easy for project managers to use a transactional leadership style on traditional projects. The triple constraints foster this style of leadership. In such scenarios, the project schedule is a goal-based motivator and can be used effectively to monitor work completed versus planned and prompt changes in work assignments when needed. On a virtual project, transactional leadership is more challenging for project managers due to the distributed nature of the team and the work. Visibility into the work and accomplishment of tasks can be quite limited, and experienced virtual project managers realize that conversations need to occur at the milestone and deliverables level with focus on team collaboration at critical integration points rather than at the more transactional, task level.

The primary areas of transactional leadership focus on:

- Consistent, standard, and well-prepared communication.
- Established and sustained roles, responsibilities, and expectations with accountability for each task.
- Regular use and communication with progress and status reports.
- Quick attention to risk trending to issues and problems.
- Individual follow-up on tasks and accomplishments.

Transformational Leadership

Bernard M. Bass extended Burns's earlier work, focusing on the transformational leadership style. Under this approach, as the word *transformation* implies, the leader uses various mechanisms to create a motivating climate for team members. The leader adopts a nurturing style, expressing care for individuals and utilizing ongoing communication and dialogue to identify with the project vision as a means to motivate team members to desired actions. This leadership style focuses significantly on the people side of the project and also contributes to managing and reducing resistance to change, whether the change is to processes, technology, development, or organizational restructuring.[16] When team leaders identify with and follow the transformational approach, there is considerable focus on fostering team identity, trust building, encouragement of communication, and collaboration and building the team's working relationships.[17] For these reasons, transformational leadership is quite effective for virtual team leadership.

A major point of contrast between transactional and transformational leadership is one of self-interest. Transactional leaders do not often influence beyond self-interest. This is limiting in many ways; for example, it limits individualized consideration, collaborative team building, and collective performance. Transformational leaders motivate and inspire by using more than simple pay incentives; they look to intellectually stimulate their team members, they look to establish meaning and purpose in each member's individual work and the collective work, and they aim to offer individualized consideration to each member of the team.

The primary areas of transformational leadership focus on these areas:

- Establishing team identity
- Establishing trust
- Building team ownership, relationships, and rapport
- Fostering collaboration and integration of work

Situational Leadership

Paul Hershey authored the situational approach to leadership. This approach to leadership is founded on the notion that team leaders need to adopt a leadership approach that fits the situation faced at any given point during the project. Doing this requires considerable flexibility and the ability to adapt quickly during the life of the project. Team leaders must be very sensitive to team members' cultures and be prepared to properly address any confusion and problems that they might present.[18]

As the name suggests, situational leadership is dependent on a particular set of circumstances. This style takes into account both task activity and team members. The goal is to understand work effort relative to the team members' ability (proven experience, confidence, and capability). For some members of the team, their experience is well established and aligned with their work responsibility for the project. In this situation, virtual project managers will not need to focus on task-level activities. Rather, team leaders should make sure these members are fully engaged, handing off their work appropriately, supporting the work of others, and have opportunities to further hone their skills and learn new ones. Virtual project managers will find team members with the opposite abilities as well. These members may not be as experienced and may be in roles that stretch their capabilities. Virtual project managers will need to spend more time helping these members understand their roles, expectations, and, generally, how to contribute meaningfully without slowing down the project or other team members.

The situational leadership style encourages virtual project managers to adopt an approach that is the best for each team member's experience and capability, relative to the given task of that team member. Doing this requires a high level of flexibility and ability to diagnosis current situations and act accordingly. As the project goes through its life cycle, project managers may very well need to adjust their leadership styles.

When reviewing the leadership styles, it becomes clear that no one style will fit perfectly for all project situations. Given the complex and challenging environment the team faces, virtual project manager need to adopt a leadership approach that, first of all fits their personality and makeup and that second, blends the elements of the leadership styles that enables the leaders to be as effective as possible. For example, Bass, through his work and research, indicated that leaders exhibit simultaneously both transactional and transformational leadership traits when leading teams through the life of a project. Last, it is important to also point out that many elements and traits associated with the transformational leadership style fit nicely with leading across multiple geographic sites in the virtual team environment, given the lack of ongoing human contact necessary for developing and sustaining relationships.

An Interview with Elliott Masie

Kathy Milhauser, a former global practitioner turned researcher and academician, had the opportunity to catch up with Elliott Masie and conduct an interview with him on the trends of virtual leadership. Masie is a futurist and researcher covering topics on workforce learning, business collaboration, and emerging technologies. Here's what Kathy uncovered from her interview.[19]

Maise said: "Healthy organizations that have really good approaches to leadership are more likely to be healthy in the virtual leadership world, because they are already addressing what the team needs." He continued: "You have to adapt in a virtual leadership mode, you have to become very conscious that you are always attending to multiple buckets." When Kathy asked for more details, Masie explained that the best leaders make sure that the "team is clear and aligned;" they also focus on "developmental" aspects of their team. Leaders need to "build and support the building of skills for people to do their tasks." The third element of great virtual team leadership, Masie said, is "to build and maintain the team culture, and that includes trust and communication." He concluded, "You have got to make sure that people are aligned and crisp and moving toward their targets, and you also have to make sure that they are developing and filling minuscule gaps. And then you have got to make sure that the culture is healthy, so that it's actually an environment in which people can thrive and succeed."

Adding "E" to Leadership

Without deviating from traditional team leadership where one aims to create and drive a project vision, bind organizational components together via a common purpose, and achieve a set of goals, e-leadership is fundamentally different from traditional project team leadership because it takes place in the virtual context and is mediated through the use of technology to communicate and collaborate.

E-leadership ("e" meaning electronic) is related to leadership in the information age, which is characterized by fast development of technology and a global economy where firms compete across national borders seeking out geographies where they can make a profit.[20] E-leadership is a descriptive term that includes a significant portion of the leadership accomplished in the virtual environment—it is virtual leadership. The term describes leadership that encompasses electronic technology to achieve a significant portion of the communication, collaboration, and performance of design and development work pertaining to virtual teams. E-leadership can be characterized as a means of socially influencing attitudes, feelings, thoughts, and behaviors through the use of advanced information and communication technology.[21]

The evolution of virtual team leadership and management toward an e-leadership environment is an important transition. No doubt, leaders of virtual project teams must be proficient in basic project team leadership skills and competencies. However, the ability for team leaders to apply these basic skills effectively through the use of communication and collaboration technologies becomes one of the key challenges and, if accomplished successfully, one of the critical contributors toward achieving a high degree of success in a virtual team's project performance.

Virtual team leaders, therefore, must ensure that members of the team become proficient in the selection, application, and use of available communication and collaboration tools and technologies. (See Chapter 8.) This also includes ensuring that all team members at distributed locations have access to the chosen tools for the project and receive adequate training on the use of the tools.

Skills training typically available for leadership and management does not broadly address the skills and capabilities necessary for achieving higher levels of performance in the e-leadership environment for today and in the future. This, in part, may be because the electronic environment is evolving so quickly that skills and capability development have been unable to catch up. In fact, limiting the definition of the "e" to just electronic technology may not be broad enough for the future. The "e" in the future may be better represented by *evolving* as being more in tune with the rapidly changing business and virtual environments.[22]

Becoming an Effective Virtual Team Leader

"I understand the importance of distinguishing the two roles that a project manager has," explains Jeremy Bouchard. "More important, I underestimated the importance of focusing on building my team leadership skills as much as building my project management skills."

This realization didn't occur to Bouchard until he began managing virtual projects. "Looking back, I didn't put much thought into leading my co-located teams. The team was formed, we all worked together, we communicated in meetings and hallways, and we collaborated. I had people issues and conflicts to deal with, but those were pretty easy to resolve when everyone was face-to-face and basically trusted one another.

"This was *not* the case when I took on my first virtual project. I had nothing but people issues right from the beginning, and I was trying brute-force solutions using project management methods." Fortunately for Bouchard, his manager and mentor recognized the problem quickly and began coaching him in team leadership principles. "Brent started by giving me two John Maxwell books that hammered home team leadership basics. We talked a lot about how to apply those basics in a virtual team setting where communication and leadership is done electronically. He also had me focus hard on establishing team trust. I now realize that trust is the foundation of all team leadership principles."

However, there is still more learning and work to be done. "I'm still working on adjusting my leadership style to fit particular situations that arise. I'm currently focusing my learning on recognizing when a situation requires a different style."

Assessing Virtual Team Performance

The virtual project team performance assessment is intended to be used by an organization to focus on a specific project. The primary benefit in using this assessment is tracking virtual project team performance over time to assess the continual growth in performance. This assessment can also be used at the completion of virtual projects. The results from the assessment are intended to serve as the basis for future organizational actions for virtual project team improvements.

The assessment works best when a team remains together for more than one project in order to minimize the disruptions created by losing personnel and having to re-form and rebuild teams. To broaden the breadth of understanding on the overall team performance, managers of an organization may choose to have a wider vested audience complete the assessment, including the virtual project manager, team members, project sponsor, the senior management team, and other key stakeholders.

Virtual Project Team Performance Assessment

Date of Assessment: _____

Virtual Project Team Code Name: _____

Virtual Project Manager Name: _____

Project Start Date: _____

Planned Completion Date: _____

Team Size: _____

Assessment Completed by: _____

Confidential Assessment: _____ Yes, confidential

_____ No, not confidential

Assessment Item	Yes or No	Notes for All No Responses
All team members were aligned to the project mission and objectives.		
There are proven examples of clarity in project roles and responsibilities.		
Project team norms and guidelines were in place, understood, and used.		
High levels of individual motivation and sense of purpose were achieved.		
Project planning was successfully achieved.		
Effective project monitoring was in place and utilized effectively by the project manager.		
Project schedules and milestones were achieved by the team.		
Project budgeting and cost controls were successfully managed by the team.		
The project team identified and managed risk properly.		
A high degree of trust among the project team was evident and observable.		
There are proven examples of awareness of team diversity and cultural management.		

Assessment Item	Yes or No	Notes for All No Responses
There are proven examples of effective conflict management.		
Effective project leadership capabilities (participative, empathy, supportive, mentoring, motivator, provide feedback) were exhibited by the project manager.		
The project manager dealt with resolving problems and removing barriers to team progress.		
The team proactively and positively managed project changes.		
The team communicated and collaborated effectively across all geographic locations.		
The team successfully managed and influenced the virtual stakeholders.		
Customer and client needs were effectively addressed.		
Senior leaders and other stakeholders assessed the project team as achieving the project goals.		
Lessons learned and retrospective activities were completed and the findings applied and shared with other projects.		

Findings, Key Thoughts, and Recommendations

Notes

1. Russ J. Martinelli, James Waddell, and Tim Rahschulte, *Program Management for Improved Business Results*, 2nd ed. (Hoboken, NJ: John Wiley & Sons, 2014).

2. Bill Snyder, "Rendezvous in Cyberspace," *Stanford Business Magazine* (May 2003): 20. Published by the Stanford Business School Alumni Association, www.gsb,stanford.edu/sites/gsb/files/2003May.

3. Martinelli et al., *Program Management for Improved Business Results*.

4. L. Lee-Kelly and Tim Sankey, "Global Virtual Teams for Value Creation and Project Success: A Case Study," *International Journal of Project Management* 26, no. 1 (2008): 51–62.

5. Richard Lepsinger and Darleen DeRosa, *Virtual Team Success: A Practical Guide for Working and Leading from a Distance* (Hoboken, NJ: John Wiley & Sons, 2010).

6. Gregory R. Berry, "Enhancing Effectiveness on Virtual Teams: Understanding Why Traditional Team Skills Are Insufficient," *Journal of Business Communication* 48, no. 2 (2011): 186–206.

7. Thomas Kayser, *Building Team Power: How to Unleash the Collaborative Genius of Teams for Increased Engagement, Productivity, and Results*, 2nd ed. (New York, NY: McGraw-Hill, 2011).

8. Bernhard R. Katzy and Xiafeng Ma, "A Research Note on Virtual Project Management Systems," *Proceedings of the 8th International Conference on Concurrent Enterprising*, pp. 518–522.

http://www.docfoc.com/a-research-note-on-virtual-project-management-systems-wtP9B0.

9. Jon R. Katzenbach and Douglas K. Smith, *The Wisdom of Teams: Creating the High Performance Organization* (New York, NY: McKinsey & Company, 2003).

10. Martinelli et al. *Program Management for Improved Business Results*.

11. Yael Zofi, *A Manager's Guide to Virtual Teams* (New York: NY: American Management Association, 2012).

12. Terri Kurtzberg, *Virtual Teams: Mastering Communication and Collaboration in the Digital Age* (Santa Barbara, CA: Praeger, 2014).

13. Paul Holahan, Ann Mooney, Roger Mayer, and Laura Finnerty-Paul, "Do Debates Get More Heated in Cyberspace? Team Conflict in the Virtual Environment," *Alliance for Technology Management* (Fall 2008).

14. Dennis J. Cohen and Robert J. Graham, *The Project Manager's MBA* (San Francisco, CA: Jossey-Bass, 2001).

15. James MacGregor Burns, *Leadership* (New York, NY: Harper & Row, 1978).

16. Harold Kerzner, *Project Management 2.0: Leveraging Tools, Distributed Collaboration and Metrics for Project Success* (Hoboken, NJ: John Wiley & Sons, 2015).

17. Bernard M. Bass and Ronald E. Riggio, *Transformational Leadership*, 2nd ed. (Mahwah, NJ: Lawrence Erlbaum Associates, 2006).

18. Paul Hershey, *The Situational Leader*, 4th ed. (New York, NY: Warner Books, 1985).

19. Kathy Milhauser (2011). Trends in Virtual Leadership: An Interview with Elliott Masie. In K. Milhauser (Ed.), *Distributed Team Collaboration in Organizations: Emerging Tools and Practices* (pp. 225–233). Hershey, PA: IGI Global.

20. Victor C. X. Wang and Susan K. Dennett, "Leadership in the Third Wave." In Victor C. X. Wang, ed., *The Handbook of Research on Education and Technology in a Changing Society*, pp. 504–517 (Hershey, PA: Information Science Reference, 2014).

21. Carita Lilian Snellman, "Virtual Teams: Opportunities and Challenges for E-Leaders," *Procedia—Social and Behavioral Sciences Conference Proceedings* 110 (January 2014): 1251–1261. http://www.sciencedirect.com/science/article/pii/S1877042813056127.

22. Margaret R. Lee, *Leading Virtual Project Teams: Adapting Leadership Theories and Communications Techniques to 21st Century Organizations* (Boca Raton, FL: CRC Press, 2014).

6

EMPOWERING THE PROJECT NETWORK

"I had to learn to let go." That is how Jeremy Bouchard describes one of the major personal transformations he had to experience to become a better manager of virtual projects. What Bouchard is referring to is the sharing of tasking, problem-solving, and decision-making power between the project manager and team members who are geographically and organizationally distributed. However, this is only half of what we call the duality of the virtual project manager's role. The other half is the need to provide strong centralized leadership, particularly during the forming and storming stages of team development (as described in Chapter 4). This centralized/decentralized duality of the project manager's role is in effect on all projects, traditional and virtual, but it is amplified considerably on virtual projects due to the lack of face-to-face interaction and social presence caused by geographic distribution of the project team. Also, the duality of the role brings to light many of the challenges associated with managing a virtual project that were discussed in Chapter 1.

As Jeremy Bouchard experienced directly, one of the most challenging and confusing aspects of managing a virtual project is knowing when to pull the reins in to assert more centralized and direct control and when to let the reins loose to allow more empowerment of the team. That challenge is the subject of this chapter.

Centralize First

As explained in Chapter 1, one of the primary differences between traditional projects and virtual projects is that virtual projects are established on a series of networks (organizational, technological, and human), not on physical location and direct interaction. The lack of physical interaction becomes a constraining factor in the creation of human networks on virtual projects. It falls on project managers to facilitate the creation of most virtual connections when a project is in the early stages of initiating and planning. To establish core elements of the project networks, project managers must *pull in* and facilitate early communication, collaboration, and personal interaction.

Networked Projects

The concept of the networked enterprise emerged in the 1990s with the realization that hierarchical bureaucracies were being replaced by horizontal enterprises that were enabled by the use of digital technology to connect organizational nodes.[1] In networked enterprises, components (people within the organization) are both independent of and dependent on other network components where knowledge is created and retained. This dependency becomes an *interdependency* when the components of the business network begin sharing common goals.

What has not been discussed in great detail is that the emergence of the networked enterprise has brought forth the creation of *networked projects*. Because of the proliferation of the use of digital technologies to share and create work products, nearly all projects are now *networked projects* that are based on a series of horizontal organizational

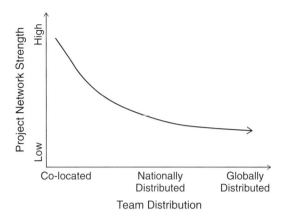

Figure 6.1: Project Network Strength as a Function of Team Distribution

connections. The strength of the project network is based on the strength of the relationships among project team members. For virtual projects, the strength of team member relationships is influenced by one key factor—the level of distribution of the project team members. (See Figure 6.1.)

Communities and Ecosystems

A couple analogies are helpful to understand the reasons behind the variation in network strength in relation to team distribution. On traditional projects, which are largely made up of co-located team members, the sharing of physical space forms a *community*. Everyone sees one another on a regular basis, some people share commonality of work by participating on the same project, and some people establish personal relationships that transcend the workplace. As a result, an organizational and project community are formed. When someone new joins the organization, they become part of the community.

In contrast, members of a virtual project do not have the benefit of sharing a common community. However, members of the virtual project *are* highly dependent on other members of the project team for their own growth and success. Instead of being part of an organizational community, a better analogy for shared dependency on a virtual project is that of an *ecosystem*.

The concept of an ecosystem was first defined in 1935 by Arthur Tansley, who characterized an ecosystem as a whole system comprised of biological organisms with a complex set of networked relationships necessary for common survival.[2] The biological ecosystem concept was then leveraged in the 1990s to define the concept of the *business ecosystem*.[3] In business ecosystems, firms are not viewed as independent entities, but rather as part of a wider network of companies that both collaborate and compete to add and extract value from the industries in which they participate. One of the most obvious business ecosystems is that which was created by Apple Computer. Networked to Apple are a large number of companies and individuals in the personal computing, telecommunications, and entertainment industries that both collaborate with and compete against Apple. The result has been unprecedented growth in the industries.

Both biological and business ecosystems provide powerful metaphors for understanding a virtual project built on a series of complex networks. As in biological ecosystems, the virtual project network is a community of agents with different characteristics, specialties, and interests who are bound together by mutual relationships as a collective whole. The fate of each agent (team members) in the ecosystem is related to the fate of the others (successfully or unsuccessfully completing their part of the project). Cooperation and collaboration between members of the ecosystem is necessary for successful completion of the virtual project.

Ecosystem Keystones

Within an ecosystem (biological, business, or project), there always exist entities that are referred to as keystone members. In complex organizational networks such as virtual projects, it is crucial that these key players or hubs exist in order to establish and enhance network stability. *Network stability* refers to the number of direct connections between the network elements. If these human hubs do not exist in a project network, the network will remain highly fragmented. On virtual projects, project managers become the primary hub, while other functional players, such as subject matter experts, team leads, and functional managers, emerge as secondary hubs. A disproportionate number of network connections flow through and surround these keystone players.

It turns out that this structure of richly connected hubs almost always emerges as networks evolve their connections over time. In a biological ecosystem, certain species emerge that serve as hubs in food webs. In project ecosystems, project managers and other keystone members act as hubs that serve as information and communication conduits between team members. (See the box titled "Connecting New Delhi and Tel Aviv.") The point we are making is that in networked structures such as virtual projects, strong hubs served by keystone players must exist to establish a high level of performance and to maintain system health.[4] Virtual project managers must realize the critical role they play as keystone hubs on virtual projects and work diligently to help establish a large number of project network connections *between* team members. Referring to Figure 6.1 again, the more a project team is separated by distance and time, the more effort project managers have to concentrate on establishing communication and collaboration connections. Even though a virtual project team may be highly distributed, early communication and integration of work must be centralized around the project manager in order to strengthen the connections within the project network.

Connecting New Delhi and Tel Aviv

For the sixth time, Abhishek tried to install the operating system software onto the new mobile communication device at his office in New Delhi, India. For the sixth time, the process failed.

The instructions for performing the installation process were uploaded to the project team's team site by Amir Matzkin, a fellow project team member located in Tel Aviv, Israel. To Abhishek, Amir seemed to be the team's expert on installing software onto the new device. After sending several unanswered emails to Amir requesting assistance, Abhishek raised the issue in the next weekly project team meeting.

Presented as an issue that would soon affect the project critical path, Abhishek stressed the need to have a direct discussion with Amir. A critical project network connection between two geographically and organizationally separated team members needed to be established.

As the project manager of the virtual project, Jason Bentley was in the best position to help establish a connection between the two team members. However, Bentley did not have a direct connection with Amir either. What to do?

Bentley assessed the project network and discovered that Amir worked in Carl Hardgrove's department. Hardgrove was a member of the project's core team and therefore, was a keystone member of the project network. A request was made to Hardgrove to facilitate the network connection between Abhishek and Amir and to ensure that a direct discussion between the two team members took place within the next three days to resolve the issue at hand.

The connection was made, the conversation took place, an error in the installation process instructions was discovered and corrected, and the software was installed on the first attempt. What should have been a two-hour task turned into a five-day delay due to a missing virtual connection.

Singular Perspective

Before a networked organization such as a virtual project can establish the interdependent connections and relationships necessary to achieve its collective goals, a change in mindset must occur. People in an interconnected network first and foremost view themselves as individuals with individual talents, skills, and needs. As we explained in earlier chapters, the *individualistic* mindset must shift to that of a *collective* viewpoint where team members see themselves as a group of individuals who share a common project mission and who must collaborate as a team to achieve project goals.

This "One Team" perspective is of course a centralized concept that must be driven by project managers. On traditional projects, this singular perspective is facilitated in large part by physical presence and social interaction. People literally see themselves as part of a team, and team identity is relatively easy to establish. This is not the case on virtual projects where social interaction and physical presence are severely limited or completely nonexistent. The "One Team" perspective must be purposefully put forth by virtual project managers and consistently reinforced from the primary hub of the project network.

Communicating Common Goals

One of the most effective ways to begin establishing the "One Team" perspective is by communicating the project goals as a shared responsibility of all team members. Centralization is therefore established through a common set of goals. The project charter (see Chapter 2) is an effective vehicle for documenting the common goals of the project. Consider the following goals statement from an example project charter.[5]

> The University is looking for new ways to help alumni stay connected to the university post-graduation. The website to be developed will 1) create the means to establish a strong alumni social network, 2) provide a portal for the university to communicate activities, information, and needs, and 3) establish a repository of academic research information for the alumni to access and contribute to. The project will be completed prior to the Fall 2019 academic semester, and cost no more than $60,000 to implement.

This simple goals statement is a very powerful mechanism for pulling a network of distributed project specialists together to achieve a common goal. What is eloquent about this goal statement is that it also contains an element of project mission:

> Creating a website that will be used as a resource to help alumni stay connected to the university and with one another post-graduation.

Additionally, this goal statement contains elements of the project and team charter. Beyond the goal itself, there is an understanding of value, success metrics, and benefit realization. A simple goal statement, when articulated well, establishes the "One Team" perspective in a compelling way.

Using Trust to Strengthen the Network

In order to begin distributing responsibility and authority to various team members on a virtual project—a necessity for effective virtual work—the connections among team members must be established and be strong. Project network connections are *established* through communication. They are *strengthened* and *maintained* through trust among project members. Trust among team members is the brass ring that binds the project network together.

Virtual teams with high levels of trust are more effective at distributing power of decision making and problem solving and are more willing to readily share information with one another.[6]

As Figure 6.2 illustrates, however, the type of virtual project has a large effect on the number of virtual network connections that must be established and maintained. The higher the degree of virtualization, the greater the number of

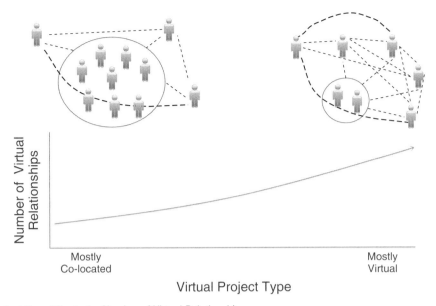

Figure 6.2: Project Type Affects the Number of Virtual Relationships

virtual connections on the project, and therefore the more tenuous the overall level of trust on the project.

Why is trust so important on a virtual project? As both authors and practitioners have observed, trust is the key ingredient necessary in preventing the geographical, organizational, and cultural distances between team members from becoming psychological distances.[7] On traditional projects and even on mostly co-located virtual projects, a high degree of physical and social presence exists, thus enabling trust to be established quickly by repeated encounters and communication exchanges. As the level of virtualization increases on a project, it becomes harder to gain the trust and confidence among team members when they never meet in a physical setting. Communication becomes less personal and more transactional.

To compensate for this, initial conversations and attempts to collaborate have to be centralized and intentionally driven by virtual project managers, along with assistance from other keystone members of the project ecosystem. In many instances, virtual team members do not even know whom to connect with when they desire to engage in knowledge-sharing and task collaboration conversations. The keystone members first establish the virtual connections and, in the process, also establish a base level of trust among members of the new connection. Once trust is established, maintaining and strengthening it becomes the next challenge.

Research from Terri Kurtzberg suggests that experts believe that maintaining and growing trust on a virtual team is almost exclusively determined in terms of reliability.[8] Specifically, virtual trust is based on two key aspects of reliability: reliability in communication and reliability in action. *Reliability in communication* refers to trusting that a response to a communication will be received from another team member when needed. Lack of response or recognition of receipt of a communication quickly diminishes trust that future communication will be responded to. *Reliability in action* simply means doing what you say you are going to do. Following through on commitments is the single most effective way to strengthen and maintain trust on a virtual project.

Establishing Commonality in Process and Tools

Virtual projects greatly benefit from commonality in methods, processes, and tools used to facilitate communication and collaboration among project team members and to integrate work outcomes across the projects. Projects are dependent on methods, processes, and tools for the success of work activities that produce the various project outcomes. Commonality in primary methods and processes drives consistency in work behavior and results. Commonality in methods, processes, and tools is best established through centralized discussions and decisions.

Virtual teams must begin with a process of convergence on which methods and processes will be common across the team and which technological tools will be adopted to facilitate those methods and processes. These decisions should not be mandated by project managers—a critical point that we cannot emphasize strongly enough. Look to the process of choosing common methods, processes, and tools as a way of building the team and strengthening the project network. Having open discussion and debate on the best methods, processes, and tools to employ will create buy-in on the part of team members. Mandating a standard set of methods, processes, and tools, by contrast, likely will serve to alienate some team members and potentially damage work relationships.

The goal of centralizing the design of the major methods and processes is to achieve consistency; however, they should not be designed so as to become overly constraining or negatively impact the natural ways of conducting project work. The classic example is trying to standardize the overall project processes to support either a waterfall or an agile methodology across the team. If the team is multidisciplinary, likely both methods (and others) can and should be used. Rather than trying to win the debate over the use of agile, waterfall, or another method, it makes more sense to standardize the methods and processes for synchronizing and integrating work outcomes, managing risk and change, and tracking and reporting project progress.

Centralizing the discussions enables making decisions on how much structure is required and how much commonality is needed. We recognize that each virtual project is unique, so what may have worked for one project may not work for another one. Factors to be considered are how distributed the team is (higher distribution usually requires more commonality), how multidiscipline the team is (the more disciplines involved, the less desire to standardize), how complex the project is (more complexity drives higher need for commonality), how many external regulations are imposed, and how many specific requirements are levied by project stakeholders.

The selection of technological tools that will be adopted and used by the project team to facilitate their virtual work is another matter that is best discussed in centralized conversations. As we describe in detail in Chapter 8, there is no ideal set of technologies for all virtual projects. All organizations are unique, as are their virtual projects and the capabilities of their project team members; so too are their communications, collaboration, and project management needs. Project technologies, therefore, must be selected based on how well they support the needs of the team, how they complement the culture of the organization and project, and how well they integrate with the suite of tools currently in use.

Creating Tacit Knowledge

The more a project team works together and collaborates on building the project plan, completing project tasks and deliverables, presenting information to stakeholders, and so on, the more they learn about one another as people and professionals. In particular, they learn who knows what, what needs to be done, who possesses what skills, how to best communicate with one another, and basic knowledge of how to work together as a team.[9]

This is known as tacit knowledge. *Tacit knowledge* refers to the things people know how to do without having to think about it. We can all think of things in which we have developed tacit knowledge—driving an automobile, playing our favorite sport, cooking a particular meal, or setting up a conference call using our company's information technology systems. After a while, we no longer have to think about how to do these things, we just do them by rote.

Creating tacit knowledge on a team is a similar process, but is dependent on how often and to what degree team members work together. Teams naturally begin to learn whom to go to for specific tasks without purposefully having to divide the tasks.[10] For example, in a closely contested game of basketball (or any sport), a team implicitly knows who the best shooters are in tense situations; therefore, players know whom to get the ball to score.

Project teams have the same characteristic, and that is built on tacit knowledge. Over time, the tacit knowledge they create guides them to know whom to go to in particular circumstances and situations.

This is another area where traditional project teams have a distinct advantage over virtual teams. Tacit knowledge is created much more rapidly and broadly on traditional teams because of the higher degree of social interaction. With social interaction limited on virtual teams, developing knowledge about such things as who knows what, what tasks need to be completed, and who are the most skilled communicators takes longer. The more work and collaboration a team can perform together, especially during the early stages of a project, the quicker tacit knowledge will be created. (See the box titled "Using Team Training to Create Tacit Knowledge.")

Using Team Training to Create Tacit Knowledge

Tanya Aquino is a colleague of Jeremy Bouchard at Sensor Dynamics. Unlike Bouchard, she was not a subject of corporate acquisition and has been with the Sensor Dynamics' parent company for nearly 10 years. For seven of those years Aquino has managed virtual projects and led virtual project teams that specialize in new product development.

Before beginning his second virtual project with Sensor Dynamics, Bouchard reached out to Aquino to see what techniques she has used to help accelerate the team-building process. "I told Jeremy that I'm a big believer in two things: bringing the core project team together face-to-face early in the project and attending some type of team-building training together as part of that initial face-to-face," explained Aquino. She added, "My favorite training is a one-day Strengths Finder workshop."

Strengths Finders is the title of a book by Tom Rath that focuses on knowing and growing your strengths, both professionally and personally.[11] "I know a wonderful facilitator that is trained in the Strengths Finder method who makes the workshop a great experience. He is a master at taking the concepts, which are really geared toward individuals, and crafting them as a team exercise," said Aquino. "What we end up with at the end of the workshop is a really good understanding of what each team member is good at, how we leverage his or her strengths, and also a better understanding of what they are *not* particularly good at. But, the real benefit for me is that we walk out of the workshop having made personal and professional connections and a strong sense of team. I believe that this investment in a single day saves me weeks of time and effort trying to pull the team together as a cohesive unit."

As Aquino explained to Bouchard, early team training is an effective method for establishing the virtual project network, for identifying the keystone members who will help connect members to other members of the distributed network, and for initiating social intimacy on the team.

Nearly all authors on the subject of virtual organizations and virtual teams discuss the importance of holding face-to-face meetings with as many team members as feasible, particularly at the beginning of a virtual project. More than anything, face-to-face meetings serve to accelerate the creation of tacit knowledge between virtually distributed team members.

Power of the Initial Face-to-Face

Raphael Sangura is an educator and industry consultant who works with companies that have moved to a virtual model for developing and producing products and services. He has consistently observed something interesting within the companies that he works with. "When I discuss the importance of face-to-face contact between members of geographically distributed teams, I get unanimous agreement from senior managers, middle managers, project managers, and team members that getting the virtual project team together is critical to developing personal relationships, trust, and distributed empowerment. However, when the virtual project managers from these firms attempt to get funding and time to bring the teams together for face-to-face meetings, their requests are often denied by the same middle managers or senior managers." When asked about this predicament, Sanguara concluded: "Either middle or senior managers don't really buy in to the importance of the face-to-face meetings, or they don't fully understand the return they will gain from their investment in money and time."

Businesses often debate, the benefits versus the costs of bringing a geographically distributed team together. However, the best virtual organizations no longer debate this issue. For them, the act of bringing their geographically distributed teams together periodically is embedded in their project planning and execution practices.

There should be no debate within your organization, either. The benefits of team face-to-face meetings are well documented in case studies,

industry research, and team member testimonials. The short list of benefits that are consistently cited include these:

- Accelerated establishment of relationships
- Increased personal bonds and trust among team members
- A clearer understanding of roles and responsibilities
- An increased commitment and accountability for meeting team deliverables and deadlines
- Broader cross-cultural awareness
- Establishment of direct lines of communication among team members
- Transfer of power from the project manager to empowered team members

When realized, these benefits move the group of individuals assigned to a virtual project to a higher degree of team performance.

These benefits are realized in large part because of an increase in social presence on a team. *Social presence* is the degree to which personal connection is established among team members. The higher the level of social presence, the stronger the personal connection is between team members. Relationships, personal bonds, trust, commitment, and team empowerment all depend on strong personal connection. Face-to-face meetings have the highest degree of social presence than any of the collaboration methods and mediums used by geographically distributed teams. (See Figure 6.3.)

As stated previously, trust is built on personal relationships. Therefore, a team cannot establish trust if its team members do not know one another. By bringing members of a virtual team together at the beginning of a project, relationships begin to form between them and personal bonds begin to strengthen. This is due to the fact that team members begin to know one another as people with personal lives, different personalities, families, and common interests outside of work. It becomes easier to trust one another when this level

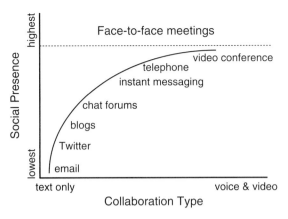

Figure 6.3: Degree of Social Presence by Collaboration Method

of understanding about one another and personal bonds are allowed to form.

With relationships and trust beginning to be established, an increased commitment to the people on the team and to the team goals begins to form. As one virtual project team member told us, "My deadlines now no longer affect a voice on the phone or a person writing an email—they now affect my friends and colleagues. I feel that my tasks are much more important to complete because I don't want to let down people I know." The implication of this is important for geographically distributed teams. By creating bonds between people through a face-to-face meeting, commitment and productivity increase rapidly.

Since personal ties degrade over time, best-practice virtual project managers have learned that periodic face-to-face meetings are required to maintain a high level of virtual team performance and commitment to project goals. This is an important concept to understand. Investing in an initial face-to-face meeting is good, but not nearly sufficient. Face-to-face meetings are necessary throughout projects, especially when conflicts arise, new members come on board, and project goals shift. All of these occurrences can begin to erode the personal bonds that are established in the initial face-to-face meeting.

The list of benefits of face-to-face meetings is long and powerful, and the value of such meetings is clear: If an enterprise wants to increase the performance and output of its virtual project teams, it must plan for and invest in periodic face-to-face meetings. This means that senior managers must provide the funding and time for face-to-face meetings, and virtual project managers must include face-to-face gatherings as part of their process and be able to facilitate the meetings effectively.

Empowering by Decentralizing

The reason project managers of virtual teams must take the team through the centralization activities described in the previous sections is to enable the team to accurately represent the interests of the project when they are empowered to solve problems, make decisions, and drive execution of work on behalf of the project manager at the local level. They must be connected through the project network to other team members regardless of physical location; they must identify with the team; they must know the goals of the project; they must have a base level of trust in their fellow team members; and they must have developed an understanding of their strengths and skills. With these actions in place, virtual project managers are in good positions to begin decentralizing tasking and execution of work, solving of problems, and decision making activities to team members who are closest to where work is being performed.

In today's nonhierarchical, networked organization, centralized command and control of work activities is rare and mostly ineffective. In networked organizations, ownership of work performance and completion of outcomes has to be distributed across the network. It is simply a case of *form follows function*—meaning that the structure of the organization influences how the work must be performed and managed. Projects, as central elements of organizations that normally mirror organizational structure, follow the same form-follows-function rule.

This means that for the networked project, project managers must delegate and grant responsibility and authority to the various team members. For virtual projects, delegation of authority becomes a risky proposition for project managers. It involves transferring an element of project leadership to another person, but to a person who is located in another geographical location, is in a different time zone, and who may have a different first language and different cultural norms. In such situations, the risk of distributed empowerment becomes higher.

To mitigate some of this risk, project managers must keep four key things in mind when delegating work virtually:

1. *Communicate the end result desired.* In order to be successful, team members who have been delegated work must have a clear understanding of what it is they are to achieve. The overall goal, scope of work, and deadlines must be clear and discussed in detail.

2. *Grant authority to work autonomously.* With the responsibilities communicated and granted, recipients of delegated work must also be given the authority to plan their work, solve problems encountered, and make relevant decisions. Project managers must explicitly communicate the level of authority they are granting to prevent recipients of delegated responsibility from having to guess their level of authority.

3. *Work, measure, and evaluate at the deliverable level.* On a virtual team, project managers will not be able to manage at the task level. Team members must have the latitude to accomplish tasks and must be trusted that the tasks will be accomplished. Project managers must "up-level" management oversight to the deliverables (and likely the major milestones) level of work.

4. *Ensure resources accompany granted responsibility.* It does no good to grant team members authority to take control of work if adequate resources are not available to perform the work. Resources include enough people, time, budget, and materials to do the job. Project managers are responsible for ensuring adequate resources are available, even if they do not personally control the resources.

This last point seems to be the most obvious factor in effective delegation of virtual work. However, it is raised most often by project team members that have been delegated additional responsibilities, as a primary reason for failing to complete the work that has been delegated. Project managers must not falsely assume that resources are going to be available to perform the work. They must continually ask those to whom they are delegating responsibility and authority if they have adequate resources to perform the work. If at any time the answer is no, project managers have the responsibility to ensure the situation is corrected, even though people within the organization who can remedy the situation directly may be halfway across the globe. If that is the case, project managers will find themselves negotiating for resources across physical, time, and possibly cultural boundaries.

Of course, delegation of work is likely a gradual process due to the underlying factor that affects project manager willingness to delegate—trust. Trust and delegated authority are closely linked, as Figure 6.4 illustrates. The more that trust increases on the part of project managers, the more likely they will be to allow team members to work independently.

Team Empowerment

In the world of networked organizations and distributed teams, the word *empowerment* is used frequently. Because of this, we seldom, if ever, stop to consider what it means. *Power*, in the context of the project environment, is the ability to effect change and influence others and having the authority to get things done.[12] *Empowerment*, then, is the act of one person sharing power with another person. It is granting another person the authority to effect change, make decisions, solve problems, set goals,

Figure 6.4: Trust and Delegated Authority Relationship

and task others on behalf of someone else—in our case, the virtual project manager.

Empowerment is the fundamental factor that makes decentralization possible on a virtual project. As team members are granted greater power, they will begin to act more on their own and rely less on the project manager's direction. They will take on greater responsibility for their work, become more comfortable with making decisions and solving problems at the local level, begin to act proactively instead of reactively to changing project conditions, and ultimately become more motivated to succeed. In their book titled *The Power of Product Platforms*, Marc Meyer and Alvin Lehnard recognize that "there is no organizational sin more demoralizing to teams than lack of empowerment."[13] This is especially true on virtual projects, which consist of distributed team members who must work independently much of the time.

Sharing of Power Takes Time

Empowerment does not, and should not, occur all at once. The release of power from project managers to virtually distributed team members must occur gradually and over time. Remember the primary theme of this chapter: Virtual projects have to be centralized first to establish commonality, then decentralization can occur over time as the team matures and trust builds. Figure 6.5 conceptually illustrates the progression of empowerment as a team moves through the various stages of the Tuckman team development model.

Why the gradual release of power? The release of power is dependent on the amount of trust and confidence that project managers have that their interests—the goals of the project team—will be adequately represented by others with whom they share their power. As we know, trust builds over

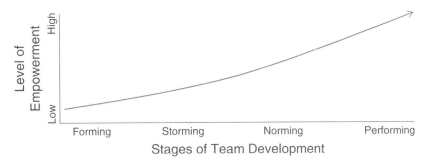

Figure 6.5: Increasing Empowerment as Team Development Progresses

time. Therefore, the sharing of power will occur over time. Even though the granting of empowerment over time can be frustrating to some team members, consider what often happens to people who have had little to no power and then suddenly find themselves in positions of influence. Many times they overcompensate, overuse their power, and get into trouble, particularly when it comes to making decisions. Other times, they simply run from their newfound power, fearing to exert their authority and influence because of the fear of making mistakes. The gradual granting of empowerment helps to ensure people are comfortable assuming authority and that they use their new power to further the interests of the project. However, project managers cannot put off granting empowerment for too long. When a project enters the execution stage, the importance of distributed empowerment increases dramatically. This puts pressure on project managers to establish trust as quickly as possible during the planning stage of the virtual project. (See Chapter 2.)

Empowerment–Risk Relationship

For virtual project managers, team empowerment is a double-edged sword. On one side, empowerment is absolutely necessary for success in a situation consisting of distributed resources. On the other side, empowerment of distributed resources means taking on greater risk. (See Figure 6.6.)

The more virtual project managers share their power, the more they are betting that the team can and will perform effectively when using that power on their own. The greater the empowerment, the less control project managers maintain.

This paradigm shift to decentralized power causes problems for some project managers as they first encounter the world of virtual projects. The natural tendency is to try to *control* risk as it emerges and increases on a project. The greater the risk, the greater the direct control project managers want to exert. Unfortunately, people who strive to assert a high level of control do not fare well as virtual project managers. Team empowerment is required

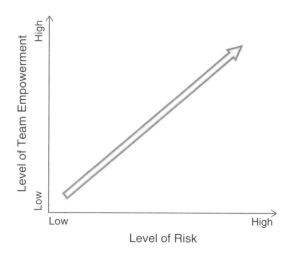

Figure 6.6: Empowerment–Risk Relationship

as well as the distribution of the responsibility to manage risk on the project.

Team member empowerment is even more critical when the project is distributed internationally. Often there is no other choice but to empower those closest to the work in order to accurately evaluate a situation and decide on the proper course of action. Slow and ineffective problem solving and decision making (which are common in decentralized models) are primary causes for schedule delays and budget overruns on virtual projects.

Because trust takes longer to develop on global projects, team empowerment is even more risky. The rate at which power is shared may therefore be slower than in domestic virtual projects. As Victor Burdic, a seasoned virtual project manager, points out, "Empowerment is something that builds over time through gaining confidence in the team." He also describes the consequence of empowerment risk: "One big screw-up on the part of someone wipes out 10 'accolades' the team has received because of the loss of trust and confidence of the project manager."

The most powerful tools for virtual project managers to use to lower the risk associated with distributed project teams are clearly defined success criteria, clearly defined deliverables and due dates, and decision boundaries based on the success criteria for each deliverable. Greater empowerment, with wider problem-solving and decision boundaries, can be granted as time progresses and trust increases regarding the team's ability to function in a decentralized manner.

Shared Decision Making

As we learned in Chapter 1, a common ailment affecting virtual teams is slow decision making. Traditional projects benefit from co-located team members who can assemble and engage in a rich discussion concerning a particular decision. Virtual teams do not have this luxury, and the separation of team members by time and distance can inhibit timely decision making. To combat this challenge, modifications to project management decision processes have to be made.

On traditional projects, project managers are the primary people providing project leadership. On virtual projects, however, leadership typically is shared among team members based on location and task at hand. This includes decision making. A more complex centralized/decentralized decision framework has to be established for virtual projects. Decisions that directly affect the success of the project—such as those that can change the project schedule, for example—need to remain centralized with project managers. Other decisions need to be decentralized and moved to where the decision outcome will be implemented. These decision types, such as the hiring of a particular project team member, become the responsibility of the project lead in the appropriate location who can effect more informed and timelier decisions. Decentralized decision making increases the probability that the right team members and stakeholders are involved in decisions and that the required data and information are available.

To make the shared decision process work on a virtual project, project managers must empower the virtual team members to whom they have delegated decision responsibility and authority. With authority granted, project managers must also communicate who has been delegated decision rights for the project to complete the empowerment process. (See the box titled "Planned Decision-Making Methodology for Maxwell's Virtual Project.") This is best documented and communicated via the team charter. (See Chapter 2.)

Planned Decision-Making Methodology for Maxwell's Virtual Project

Robert Maxwell has recently been assigned as the project manager for the design and development of a new electronic monitoring system, which will be a new product for the APEX Monitoring Corporation. APEX is a global firm that designs and develops large monitoring systems for use in industrial security systems. Robert's new project will be managed as a highly distributed virtual project due to the fact that the development team will be participating from five company sites across the globe.

Over the past few days, Robert has led a few initial meetings with the new project team to initiate his project formation activities. These sessions have been conducted as video conferences. Given Robert's prior experience with APEX, he expects that the opportunity for any face-to-face meetings with his team will be very limited and most likely any direct meetings with team members will be accomplished during visits that he plans to make periodically to each site.

Team members on this new project will be participating from these locations:

■ Los Angeles, California—Where Robert Maxwell works and will provide project leadership, as well as the location of the project's infrastructure support for administration, finance, and other support services

- Venice, Italy—Firmware, user interface, and application software. Team leader—Flavia Paola

- Cape Town, South Africa—Circuitry, power supply, and enclosure. Team leader—Joshua Vissan

- Sao Paula, Brazil—Manufacturing. Team leader—Fernando Senna

- Republic of Singapore—Quality, service, and customer support. Team leader—Jang Bahorum

Robert is now at the point in the project where he wants to discuss, clarify, and document with the team his decision making strategy for the project and how he plans to involve and empower the team to participate in project decisions. As a very experienced project manager, he has learned that for virtual teams, an appropriate level of team empowerment has worked best for his management style and enables him to achieve his best results. He has also learned that it is essential to designate who will make decisions for the team and to specify what decisions and what aspects of the decision making process must be discussed, understood by the team, and documented during the formative stage of the project in the project team charter.

Robert's decision philosophy for this project is based on a consultative decision approach. He sees decisions on the project falling into three broad categories:

1. *Project manager decisions.* Robert will be the overall project decision maker and will be responsible for delegating decision authority to others on the project. Decisions that directly affect the success of the project, as well as overall cost, schedule, and project deliverables, remain the sole responsibility of Robert.

2. *Empowerment for decision making participation by the entire team membership.* Robert has chosen to empower all members of the team to jointly participate and help decide on these team activities:

   - Establishing project expectations

   - Building team member awareness and getting to know one another

   - Developing project plans

   - Developing a collaboration methodology and deciding how the team will work together in the virtual environment

   - Establishing project team norms

3. *Empowerment of designated project team leaders.* Robert plans to specifically empower certain project team leaders to make functional expertise decisions in accordance with the project charter. These team members are:

   - Flavia Paola: Project leader, Firmware, user interface, and application software

   - Joshua Vissan: Project leader, Circuitry, power supply, and mechanical enclosure

   - Fernando Senna: Project leader, Manufacturing

   - Jang Bacharum: Project leader, Quality

These designated project leaders will be the sole decision makers for their area of expertise and will be responsible for making local decisions.

Also discussed was how decisions and problems faced were to be systematically addressed. The project manager and designated project team leaders will approach all decisions and problems faced during the life of the project by systematically:

- Identifying the problem or decision needed
- Through discussions with the project manager, identifying the specific members of the team who need to be consulted on the problem or decision needed
- Jointly characterizing the issue and documenting the problem statement and decision needed in plain language
- Identifying viable solutions and immediate and downstream implications of each
- Selecting the best choice of the potential solutions available and discussing with the team members to be consulted
- Sharing problem solution and decision with the project manager

The purpose of today's teleconference with the broader project team members is to share this plan for decision making. Once reviewed and documented, this plan will become a part of the team charter.

Decision Alignment

Delegation of team decisions, however, increases the risk that the decision outcomes can become misaligned from the goals of the project. As a safeguard, team leaders must establish boundary conditions that serve as guardrails to prevent goal misalignment. The more concisely and clearly the boundary conditions for a decision are stated, the greater the likelihood that the decision will be effective in accomplishing the direction that is needed and ensuring that the direction is consistent with the business goals driving the need for the project.

A wonderful example of a best practice for establishing team decision boundaries based on business goals is the use of a tool known as the project strike zone, which was introduced in Chapter 4. The project strike zone (shown again in Figure 6.7) is an important decision support tool that helps to keep project decisions in alignment with the business goals of the project.

The elements of the project strike zone include the business success factors for the project, target and control (threshold) limits, and a high-level status indicator. The success criteria thresholds are the decision boundaries within which the virtual team

must remain. As long as a decision will not cause the project to move outside of any of the success criteria threshold limits, the delegated decision makers are empowered to make that decision. If, however, a decision will cause the project to move outside of any success criterion (e.g., product cost), the decision must be elevated to the project manager first and then possibly to the sponsoring executive.

In practice, not all decisions can be delegated to the local level. Whenever a decision outcome will affect more than one project subteam, affect others outside of the project team, or affect the project goals, the decision must be made by project managers. In these cases, project managers must ensure that they have the appropriate information to achieve a high-quality decision outcome.

Prescriptive Decision Methodology

With decisions distributed across the virtual project, the opportunity for inefficiencies due to variation in decision methods is great. To combat this, a common decision method should be prescribed by project managers and established across the virtual team. We have repeatedly witnessed the natural desire to include distributed team members in both consensus

Project Strike Zone

Project Objectives	Strike Zone		Actual	Status
Value Proposition	Target	Threshold		
• Increase market share in product segment				
• Order growth within 6 months of launch	10%	5%	7% (est.)	Green
• Market share increase after 1 year	5%	0%	4% (est.)	
Time-to-Benefits Target:				
• Project initiation approval	1/3/2019	1/15/2019	1/4/2019	
• Business case approval	6/1/2019	6/30/2019	6/1/2019	
• Integrated plan approval	8/6/2019	8/20/2019	8/17/2019	Red
• Validation release	4/15/2020	4/30/2020	6/29/2020	
• Release to customers	7/15/2020	8/1/2020	TBD	
Resources				
• Team staffing commitments complete	6/30/2019	7/15/2019	7/1/2019	Green
• Staffing gaps	All project teams staffed at minimum level	No critical path resource gaps	Staffed	
Technology				
• Technology identification complete	4/30/2019	5/15/2019	4/28/2019	Green
• Core technology development complete	Priority 1 & 2 techs delivered @ Alpha	Priority 1 techs delivered @ Alpha	on track	
Financials				
• Program budget	100% of plan	105% of plan	101% (est.)	Yellow
• Product cost	$8500	$8900	$9100 (est.)	
• Profitability index	2.0	1.8	1.9 (est.)	

Figure 6.7: Project Strike Zone Sets Decision Boundaries

and majority rules decision methods with little success. Although the intention is correct, these decision methods are highly ineffective for virtual teams and result in significantly slow time-to-decisions.[14]

Consensus decision making strives to ensure all team members agree with what decision to make. Most of the time striving for consensus leads to exhaustive debate and can lead to circular discussions as team members toil in vain to convince others that their opinion is the right opinion. They fail to consider opening their own minds to other opinions and viewpoints. This method eventually can lead to false consensus where team members relinquish their opinion to the majority. This may be a result of having become exhausted with the circular discussions or a factor called conformity. Conformity arises in group decisions because people like to be liked. As a result, they conform to the opinions of others in order to prevent people from disliking them. It is in part for this reason we do not see "don't like" options on social media sites. We find that when either discussion exhaustion or conformity occur, disengagement once a decision is made is likely to occur.

Using a majority-rules decision method is also not recommended for virtual projects. Virtual projects often are structured in such a way that a larger number of team members are located in one or a small number of locations, with other team members distributed in smaller numbers. A majority-rules decision method can be a problem when it leads to coalition building among members who are co-located in greater numbers. Again, there is a high probability for team member disengagement in the decision process, particularly those who are distributed in smaller numbers.

The conclusion to be drawn here is that virtual project managers should avoid both consensus and majority-rules methods for decision making processes in order to accelerate time-to-decision performance and to decrease the likelihood of post-decision disengagement. Rather, the consultative decision method has proven to be highly effective for virtual projects. In consultative decisions, a single decision maker has sole responsibility for making a decision, but only after consulting with and drawing opinions from the other decision participants.[15]

Listed below are the elements required for effective consultative decision making on virtual projects:

1. Ensure there is a single decision maker.
2. Make sure all decision participants have been heard.
3. Appoint a devil's advocate to circumvent group think.
4. Give opportunity for new perspectives to be raised.
5. Ask if anyone cannot live with the decision made.
6. Strive for disagree and *commit* from those who don't agree with the decision. (See the box titled "If I Disagree, Why Do I Have to Commit?")

If I Disagree, Why Do I Have to Commit?

Debate and divergent viewpoints are good for any project team chartered with solving problems or creating opportunities for a company or a client. But, ultimately, a decision has to be made on a direction to pursue. Equally important, all members of the project team must work to execute the decision. This is where things can fall apart.

High on the list of reasons why decisions are not implemented is because the person or persons responsible for execution didn't agree with the decision and instead decided *not* to take action in support of the decision.

This is where a concept called disagree and commit can help a team. "One of the first meetings I attended at my new company," explains Jay Libby, project manager for a game console manufacturer, "consisted of a heated debate about how to resolve the high number of failures that were occurring when our new product was coming off the assembly line. Voices were raised, fists were pounded on the table, and disagreements were stated. Eventually it came down to two solutions, and a decision had to be made. At that point, the project manager acknowledged that both solutions would work and that she was going to choose solution 2 because it required fewer dependencies. Obviously, the advocates for solution 1 were not in full agreement, but then I heard the technical lead for solution 1 say, 'I disagree, but I commit.' The focus of the discussion immediately pivoted to how solution 2 could be executed, with the solution 1 advocates fully engaged in implementation planning."

This is a wonderful example of disagree and commit in action. It allows for open debate with all possibilities on the table; it allows for dissenting opinion; it provides validation that all options are heard and considered; and it provides a commitment on the part of all team members to move forward in execution of the chosen option.

What is critically important in the use of the disagree-and-commit principle is that all options and their advocates must be given a voice and true consideration. If not, the commit will never occur. Also important is a team culture that accepts the voicing of differences of opinion. This is a behavior that Jim Collins discusses in his book, *Great by Choice*. What he found on great teams is that voicing disagreement during meetings is not an option but an obligation.[16]

When all opinions are voiced, when all agreements and disagreements are discussed, and a decision is made, the team must commit to execute the decision. This principle is powerful, but if project managers are going to use it, it must be implicitly documented as part of the team norms. (See Chapter 4.)

Decision Matrix

To ensure consistency with the distributed decision making process necessary on virtual projects, some project managers use a tool called the decision matrix. A decision matrix is a table used to evaluate possible alternatives to a course of action.[17] It is used primarily to help decision makers assess and prioritize all of their options before making a final decision and setting a course of action for the project team. The decision matrix is particularly powerful when a decision maker has a number of good options to choose from. It can be used for almost all decisions where there isn't a clear and obvious preferred option. For virtual projects, it serves an additional purpose. It provides consistency in decision making methodology.

A decision matrix can be created on a sheet of paper, but it is recommended that you create a form or spreadsheet that can be reused as new decisions are encountered. Table 6.1 shows a simple decision matrix created as a spreadsheet.

To use the decision matrix, list all decision options in rows and all the relevant factors affecting the decision in columns. For example, to decide what computers to purchase for a project, factors to consider might be cost, memory size, performance, and warranty terms. Some decision makers add a weighting to each of the factors to prioritize the importance of each factor in the decision process. We recommend this approach because it creates greater numerical separation in the final scores, making the highest-priority option more obvious.

Once the matrix is set up, work down each of the columns of the matrix, scoring each decision option based on the factors. A scoring range from

0 (poor) to 5 (very good) is recommended. Next, if weightings are used, multiply the original rating for each column by the weighted rankings to get a score. Then sum all of the factors under each option. The decision option that scores the highest is the option that should be given first consideration.

The decision matrix offers four important benefits to the decision maker:

1. Understanding why and how one option is chosen over another.

2. Aligning decisions with priorities.

3. Evaluating decision alternatives on their own merit.

4. Using logic rather than emotion for decisions.

As mentioned earlier, project managers should own and maintain a centralized decision log that documents the decisions made, when they were made, and who made them. The decision log provides virtual project managers a means of keeping track of the various distributed decisions that have been made by the empowered decision makers. The decision log, therefore, becomes a key artifact in the centralized governance of a virtual project.

Staying Connected through Effective Communication

Communication is defined as the ability to transmit ideas, receive information, and interact with the environment.[18] Effective communication is crucial to the success of any project, as success is dependent on the effective transfer of information among project team members as well as with project stakeholders. This information exchange is needed to complete project tasks, solve problems, create work

Table 6.1: Example Decision Matrix									
Options	**Cost (30%)**		**Memory (20%)**		**Performance (40%)**		**Warranty (10%)**		**Score**
	Rating	Wt	Rating	Wt	Rating	Wt	Rating	Wt	
Laptop 1: Core i5	7	210	6	120	6	240	8	80	650
Laptop 2: Athlon X4	8	240	6	120	4	160	10	100	620
Pro Laptop: Core i7	**5**	**150**	**10**	**200**	**9**	**360**	**6**	**60**	**770**
Desktop: Athlon FX	4	120	5	100	8	320	6	60	600

outcomes, make good decisions, and perform all activities that are core to project work. For virtual projects, effective communication is even more than this. Since a virtual project is a networked organization, communication is the heartbeat of the project. It is first necessary to establish the connections among distributed team members, and then it keeps the connections healthy throughout the project life cycle.

Previously, we stated that trust is the foundational element that binds the project network and enables the distribution of power from the project core to key team members located where work is performed. Extending that assertion further, it is effective communication, along with consistency of behavior that maintains trusting relationships among team members and between the team and its stakeholders. As a virtual project becomes increasingly decentralized, it is communication among team members that serves the role of keeping the project network intact.

In Chapter 1, one of the primary differences between traditional and virtual projects presented was that communication is more difficult on virtual projects than on traditional projects. A simple diagram (see Figure 6.8) illustrates this challenge.

Traditional teams have the advantage of physical presence where communication can occur face-to-face. Only a portion of information exchanged between two people is accomplished through verbal exchange. The remaining information is transmitted via other means, especially through the sense of sight. It is common knowledge that when the transfer of information between

people occurs, body language adds significantly to the effectiveness of the communication.

One of the foremost researchers on the topic, Dr. Albert Mehrabian, claims that only 7% of all daily communication is accomplish by words alone, meaning a full 93% of communication is accomplished through something else.[19] Specifically, Mehrabian's studies concluded that 7% of any message is conveyed through words, 38% through certain vocal elements such as pitch and tone of voice, and 55% through nonverbal elements. (See Figure 6.9.) These nonverbal elements include facial expressions, gestures, and posture.

The exact breakdown of verbal versus nonverbal communication is irrelevant. The important point is to know that most communication is nonverbal. This is a powerful advantage for traditional project teams and a significant barrier to overcome on virtual projects. Because virtual team members cannot use nonverbal communication, they lose the ability to receive important clues that tell them messages were not fully received, were received as intended, or were not received as intended. The receiver of the messages also loses the ability to pick up on any emotions behind the sender's words.

Another factor that doesn't play in favor of the virtual project team is communication effectiveness related to the physical distance between people. Studies have shown that even traditional, co-located teams will communicate less if team members work more than 25 feet from one another.[20] Increase that distance to 100 feet, and the probability that two

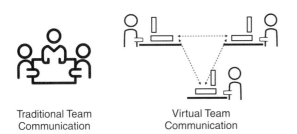

Traditional Team
Communication

Virtual Team
Communication

Figure 6.8: Traditional versus Virtual Communication

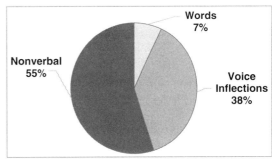

Figure 6.9: How Information Is Transferred between People

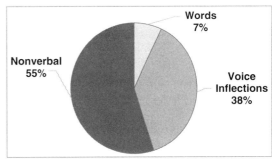
Words
7%

Nonverbal
55%

Voice
Inflections
38%

people will communicate face-to-face more than twice a week falls below 50%. The distance separating team members on virtual teams degrades communication even further if the team has to rely exclusively on written or verbal communication, and further yet, if team members do not share a common first language. (See below "Global Communication Challenges.")

Global Communication Challenges

Communication is a common challenge for leaders of virtual project teams. Although English is spoken around the globe, for most people it is a second, third, or even fourth language. Understanding this challenge starts with recognizing that communication consists of more than just language. Communication is a challenge because of frequent misunderstandings when questions, responses, and requests are not well communicated or fully understood between and among project participants. When people do not understand one another, problems flourish.

A member from our research sample, Kris Knopf, stated: "The value of communication cannot be under estimated. From my experience working across four countries, this is a constant work in process. What I have learned is that a daily phone call and email do not constitute good communication."

With co-located teams, it is common for project team members to participate in both formal and informal team meetings. Formal meetings are those held at a day and time broadcasted to team members in advance, with a set agenda and predetermined duration. Informal team meetings are those that take place arbitrarily, have no set agenda, no meeting length, and no formal invitations issued to all members ahead of the meeting time. They quite literally can be conversations that take place in the hallway or over lunch.

We find that informal meetings are just as important (or potentially more important) than formal meetings for members of a virtual team. These meetings are where much of the cross-functional communication occurs that helps to stitch together the work that spans organizational and geographical boundaries. Unfortunately, in virtual teams, these informal meetings and communications are limited to members located within the same geographical area and physical location. As a result, informal meetings between team members rarely happen, and much of the lower-level cross-functional and cross-organization communication either fails to take place or must take place via electronic means, such as instant messaging.

Becky Christopher, a manager of an international software company's program management office, explained the challenge of communicating with a globally distributed team: "As a global leader and team member, I am unable to wander down the hall to resolve challenges, momentarily brainstorm, or ask for ideas from someone who has a similar project. It can often take over 24 hours to resolve a simple question through email. When I send a question to a person in another country, perhaps 12 or 24 hours later I get a clarifying question, which is addressed by email a day later, just to find out that not all of the information is in the email that addresses my request. So I then ask the question differently—via email, of course. I finally get the data needed to address the question one or two days later." This example highlights the integrated nature of this challenge with the obstacle of time zones and emphasizes the transaction time associated with what can be a simple exchange. It can take several days to get accurate data to and from globally distributed teams as compared to several hours, or even minutes, with co-located teams.

The challenge for virtual project managers then becomes one of finding a way to compensate for the lack of informal meetings. This may mean that more formal meetings need to be set up and facilitated to promote the cross-functional and cross-organizational communication required. A delicate balance then needs to be maintained between holding more formal meetings and not subjecting the team to death by meetings.

Finally, project team size can negatively affect communication effectiveness. Not surprisingly, the larger the team, the greater the communication barriers among team members. The three most significant factors contributing to this decay in project information transfer are the fact that (1) the number of communication channels that have to be established and maintained increases; (2) the number of specialties (with specialized terminology and jargon) increases; and (3) physical distance between team members usually increases when the project is virtual.

When we add these factors to the fact that virtual project teams must communicate with one another via technology, we can clearly understand why communication is a significant challenge on virtual projects. What does all this mean? Stated simply, it means that knowledge workers working in virtual project environments have significant "overhead" associated with purposeful communication that others working in traditional project environments do not have. In addition to the focused effort required by virtual project teams to complete tasks and outcomes, significantly more time and effort must be focused on how team members communicate with one another. (See the box titled "The Five Cs of Good Communication.") Virtual communication must be purposeful, concise, and intentionally conducted on a regular basis.

The Five Cs of Good Communication

With communication characterized as the heartbeat that keeps the project network alive and functioning, it is important that communication be conducted in an effective manner across the project. These five tips by the team at MindTools provide guidance for clear communication.[21]

1. *Clear*. When preparing to communicate with someone associated with the project, be clear about what your message will convey (communication goal) and how you will convey it (messaging). Be clear about what you want the recipient to understand and keep your ideas to a minimum.

2. *Concise*. The best communications are often brief and to the point. In other words, they are concise. Strive to get your message across in a minimum number of sentences. Also strive to avoid weak words, such as "for instance," "kind of," "appropriately," or "basically."

3. *Correct*. It is unfortunate, but true, that errors in communication can easily discredit the value of a message. Take the time necessary to ensure you have not made spelling errors and have structured your written communications correctly. Be especially cognizant that names and titles are correct.

4. *Coherent*. When all points presented in a communication directly relate and are relevant to the main topic, the communication is coherent. Avoid off-topic or tangential sidebars. Coherent communication also involves using a consistent tone of messaging. Do not, for example, switch between formal verbiage and informal verbiage within the same communication. Pick one and stay with it.

5. *Courteous*. Always be respectful and professional in your communications. Strive for complete honesty and avoid talking about third parties (if they are not involved in the communication), making insults, or displaying passive-aggressive tactics. Remember two things about your communications: (1) they directly reflect on you as a person and a professional, and (2) they can resurface later to either help or hurt you.

To help ease the communication overhead, use as much synchronous communication as possible on projects. Encourage team members to use phone conversations (even if they are intercontinental) between one another instead of long email exchanges that can span many hours or potentially days. We underscore the importance of using synchronous communication for the sake of context if for no other reason. Synchronous communication allows for a continuous stream of questions and answers that provide the necessary reasoning behind a core message. This is especially important when communicating decisions or working to solve project problems.

Another important way to lower communication overhead and increase communication effectiveness on virtual projects is to establish more focused team meetings than just weekly status meetings. For example, break the typical long status update meeting into three separate meetings where performance to schedule and work accomplished is discussed in one meeting, project change management and risk management are the topic of a second meeting, and technical information exchange is the focus of the third meeting. Take advantage of video conferencing technologies whenever possible to bring in nonverbal communication. Desktop video conferencing and other tools are readily available for use on virtual projects to combine both verbal and nonverbal communication among team members and other project stakeholders. With the potential for more team meetings as a method to increase cross-team communication effectiveness, it is important to know how to conduct an effective project meeting.

Effective Team Meetings

Anyone who has had the task of conducting meetings on a virtual project will attest to the fact that they can be significantly more difficult and unproductive than meetings on a traditional project. As a result, it is common practice to attempt to replace the virtual meeting with email exchanges.

Predictably, the team learns that the time required to have sufficient back-and-forth exchange on project topics via email can be significant. Chapter 8 discusses how to be concise in writing effective emails, a good practice to adhere to. However, much of the content eliminated in emails for brevity's sake is context about why the transmitted information is important or how it pertains to the particular project situation. This results in several additional email exchanges between sender and receivers.

Instead of attempting to replace synchronous team meetings with asynchronous email exchanges, virtual project managers must focus on running effective team meetings. Effective team meetings are those that ensure adequate benefit is received in exchange for the time a meeting consumes. Team meetings should provide a forum for virtual team members to interact with one another, to discuss problems that have been encountered, to bring forth issues and potential risks that need attention, and to share ideas with one another.

Virtual team meetings have a different energy than those where people meet face-to-face around a conference room table. They have a different pace that is due in large part to the fact that work is being accomplished as team members discuss their work.[22] In many cases, work products (documents, diagrams, presentation, and so on) are created and modified during virtual meetings.

In contrast to traditional project team meetings that can be arranged quickly and often informally, team meetings on virtual projects must be more purposeful, proactively planned, and organized. As mentioned earlier, virtual team meetings must be narrower in scope and attendees. Unlike team meetings on a traditional project, where a large number of topics can be covered effectively, virtual team meetings should be limited to two or three primary topics. People get weary in virtual meetings if too much time passes and too much information is discussed. Limiting the scope will allow adequate time for additional discussion to ensure understanding and required actions.

Table 6.2: Virtual Meeting Preparation Checklist	
Status	**Checklist Items**
✓	Primary purpose of the meeting defined and communicated
✓	Expected outcomes identified
✓	Facilitator(s) identified
✓	Detailed agenda completed
✓	Start and stop times defined
✓	Presenters identified and notified
✓	All decisions to be made clearly communicated
✓	Materials sent to participants at least 24 hours prior to meeting
✓	Access to meeting technology available to all participants
✓	All participants sufficiently competent in the technology
✓	Technology is reserved, set up, and tested
✓	Participants have all access codes required
✓	All security measures identified and taken

Even with the narrower scope of virtual meetings, adequate preparation and planning is needed. See Table 6.2 for a sample virtual meeting preparation checklist. The first thing to plan is the meeting time. As Adit Liss, a seasoned project manager for Ikamai Technologies, explains, "This can be a challenge if multiple time zones are involved, especially if team members live in different parts of the world. On my previous project, I had team members in two locations in India, on both the West and East Coasts of the United States, in Germany, and in Israel. Finding a reasonably acceptable time for our weekly team meeting was tough. We finally settled on 7:00 AM on the U.S. West Coast, which was 7:30 PM in India, 10:00 AM on the East Coast of the U.S., 4:00 PM in Germany, and 5:00 PM in Israel. I feel fortunate that I am only juggling three time zones on my current project."

The next things to consider are the medium and technology that will be used—video conferencing, audio conferencing, document sharing, and so forth. With time, medium, and technology planned, basic good practices for meeting effectiveness should be followed. These include clear meeting objectives and outcomes, a specific and detailed agenda, communicated expectations to those required to present information, and all materials sent to meeting participants prior to the meeting.

For virtual meetings, it is useful to provide visual aids. Visual aids help to provide focus for the discussion and can also help the meeting participants feel more like teammates. This is particularly true if the visual aids are products of team collaboration, such as project diagrams or research findings. The use of visual aids allows for two simultaneous communication threads: one visual in nature, the other conversational.[23] See the box titled "Virtual Team Meeting Protocols" for additional guidance for facilitating a virtual team meeting.

Virtual Team Meeting Protocols

1. If work of the team is being performed in multiple time zones, rotate synchronous meetings based on time zone.
2. If the virtual meeting is an audio conference, those not speaking should mute their phones.
3. If the meeting is a video conference, participants should be aware that others can see them at all times. Side conversations and multitasking are visible.
4. Participants are responsible for giving a meeting their full attention and therefore should avoid multitasking. They should turn off cell phones, resist checking email and instant messaging, and stay with the meeting.

5. If not obvious, participants should identify themselves before making comments at all times. This is especially important in audio-only meetings.

6. A method should be used to ensure that all participants have an opportunity to participate, such as soliciting input from silent members.

7. Names should be used frequently to help participants remember who is in attendance and to assist participants to stay personally connected to the meeting.

8. The meeting facilitator should ensure that all participants understand the terminology being used. They should listen for the use of terms that may cause miscommunications, and ask for clarifications when there may be a misunderstanding.

9. All action items should be assigned an owner with clear and reasonable due dates identified.

10. Ensure that all participants can hear the conversations and can also be heard by others.

11. Participants should stay close to the microphone or camera and speak clearly.

12. The person speaking should never be interrupted, as multiple simultaneous speakers do not transmit effectively over audio and video technologies.

13. Since only about 7% of virtual communication is transmitted via words, be explicit about thoughts and feelings. Virtual meetings require explicit statements to compensate for lost communication channels.

14. Participants should avoid sarcasm, which can be misinterpreted, especially in multicultural meetings.

15. Meeting minutes should be produced and distributed within two business days following a meeting.

Once a virtual meeting is adjourned, the job of project managers as meeting facilitators is not complete. Project managers also need to attend to the conversations that continue after the formal meeting. One of the underlying goals of all virtual team meetings is to enable more communication and collaboration among team members to strengthen the project network. Project managers must be willing to participate in and facilitate follow-on conversations to ensure that this communication and collaboration is taking place.

Centralized/Decentralized Project Construct

"I remember sitting in Brent Norville's office last year when he drew a centralize/decentralize diagram on his whiteboard and explained that the diagram describes a primary construct for virtual projects," explains Jeremy Bouchard. "When I took on my first virtual project, I was lost in the diffusion of information and distribution of people and didn't understand that it was my job as the project manager to orchestrate the centralization of much of the information and then broker the distribution back out to the various distributed team members."

This was an important learning for Bouchard as he continued his transformation from traditional project manager to virtual project manager. On traditional projects, the centralized/decentralized aspects covered in this chapter occur organically and mostly by nature as a result of physical and social presence between project manager and team. On virtual projects, however, centralization and decentralization activities have to be purposefully designed, planned, and implemented by virtual project managers.

Assessing Virtual Team Collaboration

The level and effectiveness of project teamwork has a major impact not only on team performance, but also on project outcomes and the ability to achieve business goals. The Virtual Team Collaboration Assessment evaluates team success criteria that research has shown to be necessary in reaching high levels of performance.

This assessment is intended to be used by project teams and organizational managers when evaluating team effectiveness. Collaboration cannot be mandated, but it certainly can be taught, be an organizational expectation, and be intentionally designed into project work. Therefore, the results from this assessment can be used to coach and train virtual project managers and team members on good communication and collaboration practices. The assessment is meant to evaluate the project team as a whole, although reviewers of the assessment may want to use the notes section on each item to document any individual specific comments.

Virtual Team Collaboration Assessment

Date of Assessment: _____

Virtual Project Team Code Name: _____

Virtual Project Manager Name: _____

Project Start Date: _____

Planned Completion Date: _____

Team Size: _____

Assessment Completed by: _____

Confidential Assessment: _____ Yes, confidential

_____ No, not confidential

Assessment Item	Yes or No	Notes for All No Responses
The virtual project leader possessed excellent collaboration skills.		
All team members successfully participated and were actively engaged at all project locations.		
All project core team members were actively engaged in drafting the team charter.		
All project core team members were actively engaged in drafting the project mission.		
All project core team members were actively engaged in drafting the project scope.		
All project team members reviewed and understood the project business case.		

Assessment Item	Yes or No	Notes for All No Responses
All project team members were actively engaged in drafting the stakeholder list, analysis, and communication plan.		
All project team members reviewed and understood the whole solution diagram.		
Cross-team interdependencies were mapped and owners assigned across all project locations.		
There was a concise vision or goals statement for the project, and it was communicated to the team.		
Project risks were identified and discussed, and mitigation plans established.		
Honest feedback was solicited from all team members and actively provided by all.		
Project team meetings and discussions were objective, with ample time provided for questions and answers.		
All assumptions were identified and dealt with immediately.		
Any information that deviated from plans was raised proactively and resolved quickly.		
Team norms were documented and reviewed on a periodic basis.		
There was a high degree of confidence in the team's ability to succeed at all project locations.		
Each team member understood his or her role and the importance of that role in accomplishing the project's business goals.		
Team members listened to one another.		
Team members would seek to learn from each other and take every opportunity to coach and teach others on the team.		
An initial face-to-face meeting was held during the early stages of the project.		
Distributed teams had sufficient resources to conduct their work.		
Decision rights were granted to specific distributed team members.		
Decision logs were in use.		
A culture of "disagree and commit" was in place on the project.		

Assessment Item	Yes or No	Notes for All No Responses
If appropriate, team members serving as cultural liaisons were assigned to the team.		
A project communication plan was in place.		
Common collaboration tools were in place, all team members had access, and team members were trained on the use of the tools.		
Team meetings were scheduled ahead of time and could be supported in all project time zones.		
Team meetings had specified agendas.		
Proper virtual meeting etiquette was practiced in team meetings.		

Findings, Key Thoughts, and Recommendations

Notes

1. Manuel Castell, *The Rise of the Network Society*. Oxford, UK: Blackwell 1996.
2. Arthur G. Tansley, "The Use and Abuse of Vegetational Concepts and Terms." *Ecology Magazine* 16 (1935): 204–307.
3. James F. Moore, *The Death of Competition: Leadership and Strategy in the Age of Business Ecosystem* (New York, NY: HarperCollins, 1997).
4. Marco Iansiti, *The Keystone Advantage: What the New Dynamics of Business Ecosystems Mean for Strategy, Innovation, and Sustainability*. Brighton, MA: Harvard Business, 2004.
5. Russ J. Martinelli and Dragan Z. Milosevic, *The Project Management ToolBox* (Hoboken, NJ: John Wiley & Sons, 2015).
6. Terri R. Kurtzberg, *Virtual Teams: Mastering Communication and Collaboration in the Digital Age* (Santa Barbara, CA: Praeger, 2014).
7. Kurtzberg, *Virtual Teams*.
8. Ibid.
9. Ibid.
10. Tom Rath, *Strengths Finder 2.0* (Inverness, FL: Gallop Press, 2007).
11. Rath, *Strengths Finder 2.0*.
12. Jim Temme, *Team Power: How to Build and Grow Successful Teams* (Mission, KS: SkillPath, 1996).
13. Marc H. Meyer and Alvin P. Lehnerd, *The Power of Product Platforms* (New York: Free Press, 1997).
14. Lev Virine and Michael Trumper, *Project Decisions: The Art and Science* (Washington, DC: Management Concepts, 2007).
15. Chip Heath and Dan Heath, *Decisive: How to Make Better Choices in Life and Work* (Danvers, MD: Crown Business, 2013).
16. Jim Collins and Morten T. Hansen, *Great by Choice: Uncertainty, Chaos, and Luck—Why Some Thrive Despite Them All* (New York, NY: HarperBusiness, 2011).
17. www.businessdictionary.com.
18. Ed Cohen, *Leadership Without Borders: Successful Strategies from World-Class Leaders* (Hoboken, NJ: John Wiley & Sons, 2007).
19. Albert Mehrabian, *Nonverbal Communication* (Chicago, IL: Aldine Transaction, 2007).

20. Parviz F. Rad and Ginger Levin, *Achieving Project Management Success Using Virtual Teams* (Plantation, FL: J. Ross, 2003).

21. Mind Tools, "The 7 Cs of Communication: A Checklist for Clear Communication." https://www.mindtools.com/pages/article/newCS_85.htm.

22. Trina Hoefling, *Working Virtually: Managing People for Successful Virtual Teams and Organizations* (Herndon, VA: Stylus, 2003).

23. Hoefling, *Working Virtually*.

PART

IV

ORGANIZATIONAL CONSIDERATIONS

7

LEADING A MULTICULTURAL VIRTUAL TEAM

As he sits on the bullet train, Jeremy Bouchard watches the skyline of Tokyo grow smaller by the minute. It is a familiar image now, as this is his fourth trip to Japan since taking on project manager duties for the Sitka project, a project that is entirely virtual and includes an automotive partner company located in Japan. As he sits on the train taking in the sights and sounds, he realized he can even understand some of the things being said around him. The Japanese language is becoming less foreign to him.

Being able to understand some of the Japanese language is one of many ways life has changed for Bouchard since becoming a virtual project manager. It is also the result of one of the biggest changes: He now spends much more time on business travel, connecting with members of his geographically distributed team. He realizes travel is necessary to establish social presence so he is more than a voice on a conference call or a name on an email message. His name now represents a person, a person who is the leader of his project team.

Another significant benefit that has come from his travels is that Bouchard has become much more culturally aware now that there is an international component to the project. He can attribute much of that cultural awareness to one of his core team members, Takashi Kido. Kido is a Japanese expatriate who is the subcontract manager of the Japanese supplier and serves as the liaison between the two companies. As Bouchard explains, he has learned

a lot from Kido. "Takashi has stressed that trust is everything when doing business with the Japanese, and trust cannot happen until personal relationships are established. Much of my time in Japan is spent forging stronger relationships with our Japanese team members. Takashi has also explained that the organizational and management hierarchy is engrained in Japanese businesses and must be respected and used appropriately. But, the hardest cultural subtlety for me to remember is that in the Japanese business culture, 'yes' means 'I understand you.' It *does not* necessarily mean 'I agree with you.' I have had a couple instances where I assumed I had agreement on project responsibilities, only to find out later that I had assumed incorrectly."

One of the biggest breakthroughs for Bouchard has been coming to the understanding that as the leader of a multicultural project team, it is his responsibility to converge his company's culture with the national cultures represented on the team. In doing so, he is creating the project culture that is based on his company culture, but is influenced and shaped by other national cultures. Much of that influence and shaping is establishing cultural awareness and respect across the project team. His travels to Japan have been as much about teaching his Japanese team members about his company and country culture as about learning Japanese culture.

Much like Bouchard's company, Sensor Dynamics, many businesses can no longer effectively compete within their industries and markets without direct involvement by individuals located around the globe. Traditional geographic boundaries can no longer remain barriers to work progress and therefore must be managed with innovative solutions for increasingly greater and smoother permeability of what were boundaries in the past.

One of the key challenges to increased globalization is consistent success in leading multicultural project teams and interacting and integrating with fellow employees, partners, and customers from other nations and cultures in virtual team environments. Although doing this is no doubt challenging, the potential for working together and melding into teams to accomplish shared missions and goals offers tremendous benefits to all who participate.

The value and benefit of cultural diversity is well understood by leading multinational companies within all industries, as they have learned to use cultural diversity as a competitive advantage. Among the greatest benefits stated are that cultural diversity:

- Uncovers new perspectives for looking at both opportunities and problems.

- Taps knowledge and experience that is different from the members of the home country.

- Generates innovation in ideas, suggestions, and methods for performing work.

Of course, these benefits can be realized to their full extent only if members of virtual project teams become more aware of and comfortable working in culturally diverse organizations. The intent of this chapter is to help virtual project managers be prepared to operate more effectively in a variety of multicultural environments.

Putting Culture in Context

When individuals from various countries participate on a project, the influence of culture is immediate. National culture influences project team members' ability to deal with ambiguous situations and ability to work and think independently and collectively, how they view personal accountability and personal contribution, and how comfortable they are directly interfacing with people at higher levels of the organization. But what is meant by *culture*?

Culture is the total of all actions, mindsets, mentalities, and beliefs common to a group of individuals or a society. It is everything that people have, think, and do within that group or society. Culture can be exhibited in many ways, such as through spoken language, facial expressions, gestures, clothing, and rituals.[1] As displayed in Figure 7.1, many of these exhibited traits, such as language, facial expressions, and dress, are observable. However, many aspects of culture, such as values and attitudes, are not observable and reside under the surface. Therefore, multinational team members must be much more cognizant, observant, and aware in order to perceive and understand the less observable cultural attributes.

Both national and company cultures directly exert tremendous influence on us as individuals. Generally, employees master what it takes to be successful in their existing company cultures over time. The challenge facing us in today's environment of growing globalization is the necessity to expand to become more multiculturally competent to understand, manage, and lead within organizations that are composed of more employees and team members from multiple nations. This is

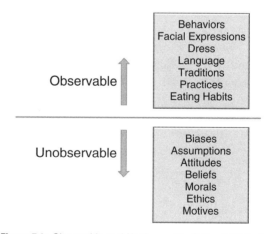

Figure 7.1: Observable and Unobservable Cultural Attributes

especially true when leading multinational virtual projects, where the collection of cultures becomes most evident and where cultural challenges first emerge. It is crucial that virtual project managers to become culturally aware and sensitive to members of their teams who represent multiple cultures and languages. Indeed, project managers in a virtual environment must pay attention to cultural differences and recognize their effect on team members' values, attitudes, and behaviors.

Cultural Intelligence

Achieving a level of cultural awareness in today's project teams means that project managers and team members understand the differences between themselves and team members from other countries and backgrounds. Cultural awareness is a subset of something much broader, called cultural intelligence. An individual who possesses cultural intelligence is skilled in and capable of understanding a culture, continuing to learn and be cognizant of ongoing cultural interactions, and has the ability to form thoughts and actions in ways that are more sensitive and sympathetic to those who represent other cultures.[2] Researchers who specialize in studying cultural intelligence indicate that it is something that we all need, and its benefits are reflected in work performance. Specifically, their studies show that a culturally diverse work group that possesses cultural intelligence will outperform homogenous teams.[3] Today it is a business necessity for virtual project managers to deal effectively with culturally diverse project teams. Project managers who do not improve their cultural intelligence skills put their virtual projects at risk.[4] This added capability includes the ability to recognize the influence of *their* culture on how they behave as leaders of project teams as well as the influence their conduct has on the team's broader behavior. Cultural difference on a project team should be embraced, not ignored, and leveraged as a source of inspiration and discovery rather than of irritation and frustration.[5] The team's diversity can be its strength.

Challenges of Multicultural Virtual Project Teams

One of the effects of increased globalization is that people may become more entrenched in their own ways of thinking, which can lead to increased divergence between cultures.[6] This behavior, of course, presents great challenges for virtual project managers. Cultural divisions will fracture the cohesiveness of the project network. As keystone members of the project network (see Chapter 6), virtual project managers must be the driving force of the convergence of cultural elements into an integrated project culture. Once again, we see the role of project managers as *integrators* emerge as a key responsibility. This responsibility adds yet another level of complexity to the virtual project, as illustrated in Figure 7.2.[7]

This complexity is magnified by the fact that culture is multilayered. Culture is like an underground river flowing through our lives—interactions and relationships providing clues; queues and messages that form our attitudes, perceptions, judgment; and ideas regarding ourselves, our team members, and others. What we see on the surface of our cultural perceptions may mask differences below the surface.[8] Virtual project managers need to be culturally knowledgeable and sensitive in managing their project team by

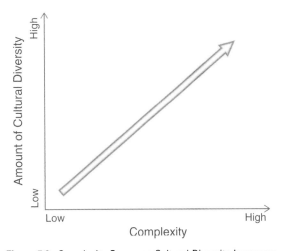

Figure 7.2: Complexity Grows as Cultural Diversity Increases

using flexibility, good communication, and mutual respect, and they must demonstrate the ability to compromise when needed.[9]

It is important to recognize that leaders of virtual project teams have in the past made mistakes, and will continue to do so as they dialogue and interact with teammates from other cultural backgrounds. This is true because, as individual team members, they are continuing to learn and practice cultural knowledge and awareness. No matter how hard team members try, they will never have complete knowledge of all of the cultures represented on virtual project teams.

Becoming aware of and sensitive to the cultural nuances that affect a multinational virtual project is necessary and foundational to effectively leading a culturally diverse project team. But, what virtual project managers ultimately have to do is create a project culture that converges both company and country cultures into a unique culture that defines how project team members will interact with one another and how the team will conduct its work. This is not an easy task, and it is yet another example of how managing a virtual project can be more complicated than managing a traditional project.

To create a project culture that takes into account cultural nuances, virtual project managers should become knowledgeable of the cultural factors that may affect their teams. That is, increase their cultural intelligence. The next section details the most prevalent cultural factors that will likely need to be taken into account when creating a multicultural project culture.

Cultural Factors

Several researchers have done a masterful job of identifying cultural factors that contribute to the cultural complexity associated with multicultural project teams. Their work is beneficial in helping project managers increase their understanding of how to recognize those cultural factors and use them to successfully manage multicultural teams. Foremost among these researchers is Geert

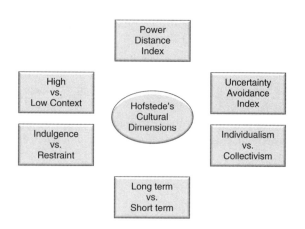

Figure 7.3: Hofstede's Cultural Dimensions Model

Hofstede, whose cultural dimensions model (see Figure 7.3) includes the cultural factors of power distance, uncertainty avoidance, individualism versus collectivism, long-term versus short-term orientation, indulgence versus restraint, and high-versus low-context cultures.[10]

Power Distance Index

The power distance index refers to how power is distributed within an organization and the extent to which the less powerful and influential individuals within an organization accept that power is distributed unequally. For example, in low-power-distance cultures, people relate to one another more as equals regardless of formal position within an organization. Subordinates may expect to be consulted by management and expect the organization to be more participative in general. Title and status tend to be less critical in lower-power-distance countries. In high-power-distance cultures, in contrast, people accept power relationships that are more hierarchical and autocratic. Subordinates expect to be told what to do, and decisions made by individuals in higher positions are rarely questioned and critiqued. Power distance tends to be higher in Eastern Europe, Asia, Africa, and Latin countries and lower in English-speaking and Western countries.[11]

Relative to virtual project teams, the power distance cultural factor could impact team members' perspectives on and understanding of the virtual project manager's management style and the communication dynamics within the company and on the project team. It therefore is important that, during discussions with the team regarding roles and responsibilities, project managers clearly specify exactly what will be the role of the team leader and what team members should expect from the leader. Project managers also should keep in mind that team members from low-power-distance cultures may want more involvement and consultation regarding potential changes and decisions on the project. In contrast, team members from high-power-distance cultures will wait to be tasked. Virtual project managers can use this knowledge to drive an appropriate level of collaboration and participation within the team. Knowledgeable virtual project managers are also more perceptive in sensing power distance discomfort and frustration within the team and respond accordingly. (See the box titled "Differing Power Distance Orientations.")

Differing Power Distance Orientations

High Power Distance (Hierarchical Orientation)	Low Power Distance (Participative Orientation)
Use senior management to make announcements and communicate changes.	Include team members in discussions and explain reasoning for directions and decisions.
Use formal management power to exercise authority.	Enable team members to ask questions and to challenge as appropriate.
Inform team members what to do instead of leaving them to figure it out on their own.	Provide a forum for team member discussions on project-related topics.[12]

Uncertainty Avoidance Index

The uncertainty avoidance index reflects a culture's tolerance for ambiguity, and it is a good indicator of whether individuals feel comfortable in unstructured situations. Team members from high-uncertainty-avoidance cultures are less likely to be comfortable in roles that are poorly defined or have ambiguous goals. In general, people from such cultures judge what is different as dangerous and risky. Team members from high-uncertainty-avoidance cultures may also seek more detail on project plans and expected outcomes and may prefer more predictable routines. High-uncertainty-avoidance cultures can be found in Greece, Yugoslavia, Japan, Belgium, France, South Korea, Italy, and many Latin American countries.

By contrast, team members from low-uncertainty-avoidance cultures are more comfortable in ambiguous situations and are more willing to take higher levels of risk to achieve an outcome. In general, team members from such cultures view new things with curiosity and seek opinions that may be in conflict with their own. Low-uncertainty-cultures can be found in North America, Great Britain, Hong Kong, Singapore, Sweden, Norway, Malaysia, and India.[13]

Virtual project managers have to be careful to match team assignments to individuals possessing the appropriate level of uncertainty avoidance. Additionally, project managers should ensure that team members from high-uncertainty-avoidance cultures have access to project requirements, plans, and schedules and have clear expectations on their roles and responsibilities.[14] Change must be tolerated, but carefully managed to allow necessary change to happen and to assure team members that the change is not occurring in an ad hoc manner. (See the box titled "Dealing with Certainty and Uncertainty.")

Dealing with Certainty and Uncertainty

Need for Certainty	Tolerance for Uncertainty
Provide team members with specific tasks, expected outcomes, and goals.	Reward team members' creative behavior and willingness to take risks.
Recognize team members' need for project information.	Focus on the process of team learning.
Focus on compliance with project processes and procedures.	Relate issues and inquiries to the project mission and goals.
Use formal communication channels.	Enable team members to challenge and question the way things are done.[15]

Individualism versus Collectivism

Individualism versus collectivism addresses the degree to which people prefer to act as individuals rather than as members of a collective group. It also generally represents the degree to which individuals in a society want to become integrated into groups and other relationships. Individualism is more prevalent in North America, Great Britain, Australia, Italy, France, and Germany. People from high-individualism cultures are comfortable working independently, generally maintain looser ties to groups, and value individual recognition as much as team or group recognition.

People from high-collectivism cultures tend to be integrated into groups and value a strong identity with the group. They put the needs of the group before their own and stress the need for belonging and the maintenance of group harmony. Collectivism is strong in Central and South American countries and most Asian countries.[16]

Relative to virtual projects, team members from collective cultures tend to expect more interaction and involvement in relationships than team members from cultures strong in individualism. Virtual project managers should keep in mind that team members from collective cultures generally are

not comfortable being singled out for recognition. Whenever possible, project managers should attempt to get to know the team members personally and try to build relationships beyond the work environment. Obviously doing this will require some travel on the part of virtual project managers. Highly individualistic members of a project team need to remain aware that they are members of a broader team and cannot perform all work independently. Virtual project managers need to be cognizant of the fact that team members' expectations about team unity, differences in personal bonding, and potentially the ways rewards and recognition are provided may be critical to the team's cohesiveness and should be subtly assessed as to what approaches work best in these situations. (See the box titled "Interacting with Individualistic and Collectivistic Team Members.")[17]

Interacting with Individualistic and Collectivistic Team Members

Individualistic Orientation	Collectivistic Orientation
Encourage team members to ask questions and seek information.	Focus on how change is good for the team as a whole.
Appeal to their observed self-interest when assigning tasks.	Appeal to their desire to work in groups when assigning tasks.
Provide individual awards when warranted.	Allow the small groups to spend time working out issues.
	Provide team or small-group awards when warranted.[18]

Long-Term versus Short-Term Orientation

Long-versus short-term orientation describes a society's time horizon and the level of importance that is placed on past, present, and future events.

It also reflects whether members of a culture prefer immediate fulfillment of their material, social, and emotional needs or if delays in this fulfillment are acceptable. In cultures with long-term orientations, values include perseverance, thrift, ordering relationships by status, and reflecting a sense of humility. Cultures with short-term orientations tend to be adaptable to changing circumstances and also emphasize quick results, personal stability, and sacrosanct traditions. In business, cultures that are short-term oriented value bottom-line results, and managers are consistently evaluated on their ability to deliver the results. East Asian countries scored the highest in long-term cultural orientation, while the United States, Australia, Latin America, Africa, and Muslim countries tend to have short-term orientations.[19]

On a multicultural project team, individuals from cultures with short-term orientations will most likely need less encouragement in holding to schedules as they tend to receive satisfaction from quick accomplishment and results on their tasks and assignments. In contrast, project managers can expect team members from cultures with long-term orientations to be more flexible and adaptable to project changes and adjustments. Project managers should ensure that team members from these cultures are encouraged to participate in project discussions relating to longer-term objectives and expectations.

High-Context versus Low-Context Cultures

Two key authors, Geert Hofstede and Edward Hall, independently developed paradigms for identifying cultural aspects which included high-context versus low-context cultures. The anthropologist Edward T. Hall in his book *Beyond Culture*, wrote about high-context and low-context cultures in regard to communication. *Context*, as used here, refers to how individuals in various cultures prefer to share information with others. It can relate to elements surrounding the communication process, such as tone of voice, facial expressions, and body language.[20]

Individuals from high-context cultures normally have extensive information networks and need a minimum amount of background perspective and data to go about their work. Generally, agreements are reinforced through their personal relationships.[21] Individuals from high-context societies generally prefer more specific information than individuals from lower-context cultures, and they tend to focus on building harmony and relationships with others. They generally speak in an indirect manner, avoiding speaking negatively of others as much as possible. Higher-context individuals tend to thrive in professions such as marketing and human resources.

Individuals from lower-context cultures tend to be more task oriented. Such individuals might prefer working in professions such as engineering and finance. Typical examples of low-context societies are the United States, Canada, Israel, and the majority of northern European countries. Chinese, Japanese, African, and Arabic societies are examples of high-context cultures.[22]

Team members from high-context cultures might not understand communications such as emails and other forms of messaging well without more information provided on the topic or context. Members from lower-context cultures focus more on objective and factual information and require less background information. Virtual project managers should keep in mind that team members from low-context cultures may prefer more asynchronous communications whereas high-context personnel will seek out information-rich communication and collaboration channels such as video conferencing.

Interactions among low-context and high-context individuals may be somewhat challenging in virtual project teams. Therefore, the project managers need to remain cognizant of these potentially conflicting situations. As an example, high-context individuals may need more data and information pertaining to the project and specifically to their own assignment and responsibilities. Team members from low-context cultures may perceive being asked to provide data or information as insulting, as evidence that other team members do not trust them, or that others "just don't get it." This fact leaves significant opportunity for verbal and written communications to be misinterpreted.[23]

Hans-Juergen Junkersdorf, senior business development manager for Motorola, had this to say about leading a multinational project team consisting of individuals from both low- and high-context cultures: "When leading a cross-cultural virtual team, you must know and understand the basics and the differences of the various cultures you will be working with on your team. The key is being able to adjust your leadership to fit the needs of both the high and the low context team members." He added, "Team members from both extremes can exhibit hidden agendas which could impact your project. Hence, the project manager needs to listen carefully and make an effort to read the 'white spaces' between their comments and actions and combine that with the facts to gauge how best to adjust your leadership approach and how to manage the flow of project information and communication."

This is very good advice from someone who has spent a number of years leading virtual project teams made up of members from multiple cultures.

Indulgence versus Restraint

Always looking to improve his model, in 2010 Hofstede added a sixth dimension described as indulgence versus restraint. Indulgence reflects a society that values relatively free and open gratification of basic and human impulses and desires. It pertains to enjoying life and having fun. Restraint, in contrast, relates to a society that values limitations and control of gratification of needs and attempts to regulate it through strict social norms. For example, such things as freedom of speech and the desire for significant leisure time are considered very important to individuals from societies that are high on the indulgence dimension. According to Hofstede, indulgence societies include North and South America, Western Europe, and parts of Sub-Saharan Africa.

Examples representing norms that are high on the restraint scale include stricter sexual norms, a perception of limited ability to affect the actions of a higher power, and lower emphasis on leisure time. Restraint is more prevalent in Eastern Europe and Asia and in the Muslim world. Mediterranean Europe tends to be middle of the road on this dimension.[24]

For project teams working in a multinational environment, this dimension may have influence on how willing team members are to express their opinions and provide feedback, especially if some team members are from a culture higher on the restraint scale. It also may have a bearing on turnover of team members and employees if they place considerable value on personal happiness, freedom, and leisure time. They may become uncomfortable in their current role and choose to explore new opportunities.

The cultural dimensions discussed above have predominantly been associated with the research work of Geert Hofstede. There are other cultural dimensions that have been identified that have a significant bearing on multinational virtual project teams that include convergent/divergent tendencies, assertiveness and language proficiency.

Convergent/Divergent Tendencies

Another cultural factor that may come into play on a multinational virtual project is known as *convergent/divergent tendencies*. Individuals from convergent societies have a preference for working on well-defined tasks and problems. They like to gain knowledge through learning while doing. They lean more toward being pragmatists who are more concerned with what works than with what appears to be true.

In contrast, people from divergent societies tend to prefer to work on vague and less defined challenges where many alternatives exist for a solution. Individuals from these cultures use more creative and holistic approaches to perform their work and concentrate on observation as a means of learning. Based on the ability to conceptualize, many researchers and authors associate Americans as convergent thinkers, whereas they see Europeans as more divergent thinkers focusing on larger concepts, such as from what philosophical position is the proposition derived.

It is highly unlikely that either convergent or divergent thinking alone can be used to lead virtual projects or solve problems related to them. For example, during the initiation phase of a virtual project, the team needs divergent thinking to analyze and create viable possibilities and solutions. During the planning and execution phases, the team needs convergent thinkers to pragmatically select the best project options and rally team members toward execution and implementation.[25]

Assertiveness

Assertiveness is a rather tricky cultural factor to manage on virtual projects. Assertive behavior is generally associated with goal-oriented individuals. Studies have revealed that males tend to be more assertive than females. However, a more persuasive cultural consideration may relate to our earlier discussion regarding individualist versus group orientation. Research indicates that participants from individualist cultures are more assertive than those from group oriented societies. One explanation given for this observation is that certain levels of assertiveness may not be tolerated in some collective societies. Variations have been observed in the degree of assertiveness in specific cultures, but it may not be appropriate

to assume that East Asians, for example, are less assertive than North Americans. It is possible and likely that people from East Asian and other similar cultures may exhibit assertiveness in certain situations, but not in others. Studies have also found that individual attitudes versus social norms may be a key contributing element that distinguishes assertive cultural behavior from nonassertive behavior.[26]

When leading virtual project teams, team member assertiveness may be perceived as an asset or liability. Different cultures vary in their perceptions of some cultural factors, and assertiveness is one of the most divisive. Virtual project managers must remain cognizant that an individualistic approach of being brutally honest with a team member versus the collectivistic approach of face-saving may resonate well with individuals from some cultures, but be offensive to others. Therefore, project managers must be able to manage assertive team members who may reduce involvement and participation of team members from cultures who react negatively to assertive behavior.[27]

Paola Genovese, Marketing Director and experienced virtual program manager with Cinetix, views her assertiveness as a major asset. "Assertiveness is a key capability of a virtual project manager. Project managers are usually managing resources which are not under his or her direct control, meaning that the project manager is not the direct manager of team members. For this reason assertiveness, and in particular the capability to use assertive communication, is critical to managing the team. Leveraging good assertive communication, the virtual project manager can improve his or her leadership of the team. This is accomplished through providing balance and effectiveness and consistently working in the same direction as a team, no matter how distributed the team is, in order to successfully achieve the project goals."

Language Proficiency

It has been stated that English (or, broken English) is the language of business. However, the English language is one of the most challenging languages to translate into native languages and vice versa. Many English words have multiple meanings, and a common expression in English may have a different meaning in other languages. Most members of multinational virtual projects are assigned to the projects because of their innate technical skills, rather than their language skills. In such cases, much can get lost in the communication process through improper translation. As illustrated in the Telephone Game in Chapter 1, disconnects in communication transfer regularly occur even among people who are co-located, from the same country, and speaking the same language. These language disconnects are magnified with virtual teams from multiple countries speaking multiple languages. These disconnects can create a loss of information and misinterpretation of critical project information during meetings and other forms of communication. And, of course, more is lost due to the lack of visual cues available during face-to-face communication, which is normally limited or not available to virtual teams, as discussed in Chapter 6.

It should be recognized, and team members should be sensitive to the fact, that people working in their non-native language can be at distinct disadvantages. It may take longer for them to communicate to adequately ensure that they are getting their messages across and that they fully understand what is being said to them and around them in the project team environment.[28] Ideas that come from non-native speaking team members often are interpreted as more simplistic. Without adequate sensitivity, these team members are often afforded less respect and credit for their contributions.

There is no doubt that language proficiency is a valued and important skill for virtual project managers and members of internationally distributed teams. The need for language proficiency will continue to grow over time as more project work is distributed across the globe. Even today, many teams may conduct the majority of team meetings in English, but in practical terms, many other languages are used in communications specific to particular regions represented on the project. As this situation continues to increase, more and more project teams are realizing the need to audio record all meetings so that participants can listen to the parts of the meeting again as necessary. Companies are also recognizing the need to invest in language training for virtual team members, especially in the predominant language of virtual projects.

Evaluating Western versus Non-Western Culture

We wrap up our discussion on cultural factors by categorizing cultural factors in a manner often described as Western versus non-Western countries. Table 7.1 displays various cultural factors associated with this type of categorization. The side-by-side comparison is meant to provide a snapshot comparison of Non-Western cultures to those in the West. It is important to remember that these comparisons are generalized and may be good for a quick, initial assessment, but a more thorough analysis should be conducted to get a clear sense of similarities and differences in the cultural values represented by virtual project team members and stakeholders. This table provides a broader perspective based upon global regions on groupings of cultural observations that virtual project team members may come in contact with during their project assignments.

Table 7.1: Western versus Non-Western Cultural Factor Categorization	
Western Cultural Values	**Non-Western Cultural Values**
Low power distance	High power distance
Low uncertainty avoidance	High uncertainty avoidance
Individualism	Collectivism
Short-term orientation	Long-term orientation
Indulgent tendency	Restraint tendency
Low context	High context
Personal achievement	Collective achievement
Equality	Hierarchy
Self-interest	Saving face
Respect for results	Respect for status
Personal rights	Social responsibility
Assertive and direct	Humility and passive
Self-assuredness	Self-abnegation
Vitality of youth cherished	Wisdom of years cherished
Materialistic	Spiritualistic
Seek change	Accept what is
Freedom of speech	Freedom of silence
Wealth viewed as result of enterprise	Wealth viewed as result of future
Outer-world dependent	Inner-world dependent

Creating a Cultural Strategy

Gaining an awareness of the cultural factors that are present on a multinational project is good practice that creates cultural intelligence for virtual project managers. But, that knowledge is of little value unless it is put to use toward leading the project team more effectively. The best project managers go beyond gaining an awareness of the cultural dichotomies on a team. They use the information and knowledge gained to reconcile those cultural dichotomies through modifications to how they lead team members and how they manage team communication, collaboration, andv decision making. To do this, project managers must develop a cultural strategy that reflects the cultural nuances and dichotomies on their team and which assists them in determining how best to modify their leadership and management approach in a multicultural project environment.

Cultural Self-Awareness

The first critical step in developing a cultural strategy is to become grounded in the cultural factors that will likely be present on a project based on the countries that are represented. This is where the cultural factors information detailed in Table 7.1 comes into play.

Normally, firms that perform project work internationally do so on a regular basis and have well-established sites and partners in various countries and regions. The national cultures represented on the company's projects are therefore mostly known prior to beginning the projects. Savvy virtual project managers will prepare in advance of a project assignment by studying the unique cultural factors that they will encounter on the team based on the countries in which their company operates. The project managers will then further refine this knowledge and understanding once their specific project team members are assigned. If possible, they may also take advantage of cultural awareness training and publications to learn more about specific cultures.

Since people are a product of their own culture, including project managers of multinational projects, we need to increase our self-awareness of cross-cultural values and behaviors in order to better understand one another. Project managers must be the champion for cross-cultural leadership and model good culturally sensitive behavior while leading project teams. (See the box titled "Seven Cultural Behaviors to Model.")

Seven Cultural Behaviors to Model

Stephanie Quappe, founder of Inter-Cultural Change Management Organization, and Cantafore Giovanna, consultant, suggest six useful behaviors for multinational project managers to personally demonstrate in order to increase cultural sensitivity on their project teams.[29]

1. *Admit that we don't know everything.* Much of our current knowledge regarding other cultures may be incorrect or biased based on stereotypes.

2. *Suspend judgments.* Do not make generalized or stereotypical assumptions about team member's cultures.

3. *Show empathy.* By listening and caring about others, we learn how other people would like to be treated.

4. *Systematically check our assumptions.* Ask for feedback and constantly make sure you clearly understand a situation.

5. *Become comfortable with ambiguity.* Accept the fact that internationally distributed teams will be more complex and that many things will not be completely clear.

6. *Celebrate team diversity.* Recognize and espouse the value of differing viewpoints, opinions, and ways of doing things on the project.

Once project managers are assigned to a virtual project, they will gain specific information on who will be involved on the project and where they reside. This information allows project managers to move to the next step in building a cultural strategy: assessing the specific cross-cultural factors that are present on the project.

Assessing the Project Team Culture

Performing a cultural assessment based on national traits may seem too general and a bit stereotypical to some, but the intent is to make the major cultural dichotomies present on a multinational team visible and clear. Within any national culture, individuals vary. Therefore, we recommend performing the assessment on a national basis.

The cultural assessment is a useful tool for virtual project managers to evaluate the varying cultural dimensions present on their teams.[30] Cultural assessments are valuable in evaluating how a new multinational project stands with respect to cultural factors. But they can become quite complex and require an inordinate amount of time to generate. We advocate a simple approach that may be less precise than other approaches, but provides all the information necessary to develop a cultural strategy for the project. Figure 7.4 is an example of a simple

Cultural Dimension	Sensitivity Low ← → High				
	1	2	3	4	5
Japanese Team					
Power Distance				X	
Uncertainty Avoidance				X	
Individualism	X				
Long-Term Orientation				X	
English Proficiency			X		
Assertiveness	X				
American Team					
Power Distance	X				
Uncertainty Avoidance		X			
Individualism					X
Long-Term Orientation		X			
English Proficiency					X
Assertiveness				X	
Chinese Team					
Power Distance					X
Uncertainty Avoidance			X		
Individualism			X		
Long-Term Orientation					X
English Proficiency	X				
Assertiveness			X		

Figure 7.4: Cultural Assessment Example

cultural assessment tool that can be implemented easily.

The structure of the cultural assessment tool must be customized for the organization in which it is used and possibly for specific projects. The two primary structural components that need customization are the specific national cultures that will be evaluated and the cultural dimensions chosen as most important to evaluate. The teams represented should be straightforward based on the geographic locations of a multinational company. However, at times team composition may have to be modified. A good example of this comes from our friend Jeremy Bouchard, who has to include a Japanese automotive partner that is specific to the project he is managing and therefore must include the Japanese culture in his assessment.

Defining the appropriate cultural dimensions requires careful thought. This is where project managers' cultural self-awareness comes into focus. The overall intent of the cultural assessment is to identify any cultural extremes that may affect project management and collaboration among project members. Therefore, it is important to select cultural dimensions that will make the cultural extremes among the participating countries visible. For example, in the assessment example shown in Figure 7.4, one North American and two Asian countries are represented. Project managers with some cultural awareness knowledge will understand that there is likely to be some variation in cultural factors. We recommend using no more than six or seven dimensions when structuring a cultural assessment.

Once they construct the tool, virtual project managers can use it to assess the sensitivity to each of the cultural factors that affect their teams. To emphasize a point made previously, the sensitivities will vary from person to person on the team, so it is best to evaluate the team from a national perspective, not an individual perspective. For the example shown, the project manager has made an assessment that the team members from Japan will be relatively low on the individualism cultural factor. The manager must then determine just how low on the sensitivity scale the Japanese team members are relative to the rest of the team. There is no quantitative approach to this. Qualitative assessment skills must be used to make these determinations. A relatively accurate assessment of a team's cultural sensitivity can best be gained by personal and direct interaction with specific team members, just as Jeremy Bouchard did by traveling to meet his new Japanese team members.

The key to using the cultural assessment effectively is to identify the cultural factor extremes (either very low sensitivity or very high sensitivity). After identifying the extremes, project managers can develop a cultural strategy and use it to make adjustments to the way they manage the project and lead the team. These strategies can also be used to train and develop project team members and help navigate team members from forming to high performing.

Cultural Strategy

With the information contained in the cultural assessment, project managers can determine if there are any cultural circumstances unique to the project that have to be addressed. As an example, Figure 7.4 shows that there are significant cultural differences in the areas of power distance, assertiveness, language proficiency, and individualism that need to be addressed. Project managers must develop a cultural strategy that addresses *how* the differences in the cultural factors will be dealt with effectively. Every multinational project team is unique, so every

cultural strategy will be unique. However, for the example shown, four cultural strategies could be developed.

Strategy 1: Power Distance

Since it was determined that both the Chinese and Japanese team members are from high-power-distance cultures, they may be sensitive to interacting with senior leaders within the company. The project manager should therefore be careful not to have these team members directly interface with or present to the senior organization leaders, at least not initially. The project manager should also ensure that clear roles and responsibilities are documented as well as all decision makers on the project.

Strategy 2: Assertiveness

The cultural assessment also showed a wide dichotomy with respect to assertiveness. This may be a source of contention even in team meetings; some participants may feel that a meeting contains lively debate while others might feel attacked or unnecessarily challenged. The project manager must monitor and facilitate the tone of cross-team communications, both verbal and written, until the variation in assertiveness begins to diminish and the team achieves a "performing" level of development. This behavior is also a team norm that should be addressed, documented, and discussed as part of the team charter. The project culture may require open and direct debate; therefore, expectations and understanding have to be established across the cultures represented on the project team.

Strategy 3: Language Proficiency

Proficiency in the primary language of the project (English, in this case) was identified as a cultural factor with high sensitivity. The team members from Shanghai especially had a low proficiency for communicating in spoken English. Due to this cultural factor, the project manager cannot rely heavily on verbal communication, but rather must establish a practice of repeating, in written form, any verbal instructions, critical discussions, and decisions. This will help to ensure that the team members in Shanghai understand and are fully engaged in the dialogue and work of the project team.

Strategy 4: Individualism versus Collectivism

Cultural preference for individualism ranges from high, to low, to moderate among the three nations represented on the example project. This distribution of preference is likely one of the most difficult factors to alleviate because it affects two major elements of any project: the tasking of work and rewards and recognition. Team members high on individualism can be tasked and rewarded individually. However, team members low on individualism and therefore high on collectivism should be given tasks that can be performed in small work groups. Rewards for these team members must be given as team awards. The virtual project manager must work hard to avoid demotivation and conflict.

With the cultural strategies established, project managers must use the strategies to modify how they will lead the people on the team and how they will manage project processes. Managers will, in effect, be using the cultural strategies to create an element of the project culture.

Converging Company and Country Culture

As an enterprise expands its business globally, it establishes operational components in a number of geographical regions and countries. The people acquired through mergers and acquisitions in each country come to the company with their own unique values, preferred behaviors, and languages that embody the culture of their society. Additionally, company acquisitions and strategic partnerships will bring new company cultures as well as national cultures into the enterprise. Managing across cultures with virtual projects requires integrating many different functional disciplines

and support group personnel who all possess obvious and non-obvious differences in backgrounds and languages. Doing this requires the ability to integrate national, company, and project culture in a way that promotes collaboration and collective thinking. According to Margaret Lee, author of *Leading Virtual Project Teams*, "Successful integration of national cultures into one strong organizational culture can be an advantage in the competitive global marketplace. Ample evidence supports the fact that integrating organizational and national cultures is necessary for organizational effectiveness in the 21st century."[31] How are the many cultural differences reconciled to the point where a company remains operationally effective and maintains its original culture?

Authors and researchers continue to debate whether company or country culture has a stronger influence on team members. The truth of the matter is that it most likely depends on a host of many varied yet interrelated factors. Companies should comprehensively assess their organizational culture against the various local cultures, countries, and regions where they do business. It is not unusual for potential conflicts between organizational and national cultures to surface. Companies should be cognizant of these situations and plan to take the appropriate actions to ensure that employees from other countries remain motivated and committed to the organizational mission and objectives.[32]

Some people talk about the need to blend the cultures into a new culture that encompasses elements of each country's culture. In practice, doing so is a nearly impossible task, and as Mary Dunkin, a global project manager for Intel Corporation, explains, in cases where a company's culture is one of its competitive strengths, creating a blended culture can be counterproductive.[33]

> In the early 2000s, Intel decided to enter the communication industry to augment its computing business, and did so through a significant number of company acquisitions. Over an 18-month period, we acquired over 20 companies. Over half of them were

outside of our home country. In nearly every acquisition, the approach was to allow the company to remain autonomous in its operation and culture. Looking back, this was a poor approach and found to be the root cause of Intel failing to retain the value that they originally saw in these companies. By allowing the companies to operate autonomously operationally and culturally, we were not able to leverage Intel's vast assets and strong company culture that are both competitive advantages.

Dunkin then went on to explain, "we then tried a blended approach, particularly in regard to cross-country culture. A lot of time, effort, and money was expended on cultural sensitivity training and coaching. The goal of this program was to increase cultural sensitivity on the part of the Intel team. Also, to help blend the Intel and U.S. cultures into the acquired companies, we assigned a small senior leadership team to work in-country and alongside the acquired leadership team. Despite this blended approach, the situation became worse. The strong company culture of Intel began to fragment and dilute to the point where it became less of a competitive advantage." Dunkin concluded from her extensive experience: "Cultural sensitivity is necessary, but only to a certain point. The point where it becomes counterproductive is when it begins to negatively affect a company's base culture. Company culture has to remain dominant over country cultures."

Recent research supports Dunkin's opinion. It suggests that a strong company culture serves a utilitarian purpose as it tends to enable setting expectations, increasing the likelihood that when members of a team and other employees are faced with uncertainty, they will act in a consistent manner. A strong corporate culture sets the rules of engagement. Researchers have further demonstrated that within a group or team setting, when coordination is rewarded and when the group's or the team's culture is strong, its members become increasingly like-minded.[34]

The benefit of cultivating a robust and solid company culture is that it enables the establishment of common values and aligns employee behaviors. Multinational corporations use many means and avenues to build company culture, such as employee handouts, corporate ethics guidelines, written value definitions, new employee on boarding activities, and worldwide company meetings with all employees to instill and align employee values and behaviors. Additionally, in some multinational corporations, targeted hiring and employee self-selection leads to employees at company sites in other countries to become more in harmony with the respective corporate culture. Those employees who fit well stay with the company, and those who do not either do not get hired or leave after a period of time. Companies that nurture this approach may be able to maintain a consistent culture across their international locations.[35]

Converging Cultures

Rather than trying to blend cultures in a multinational organization and on a multinational project, it may be better to think in terms of converging the various cultures. Three primary steps are involved in the convergence of cultures:

1. Establish a dominant cultural model where one ideology and set of behaviors is declared the official way of doing business.

2. Establish an integration of team identity model that drives a common set of motives, ideas, values, and goals that members of the organization can all relate to.

3. Maintain cultural awareness to recognize and respect the differences each culture brings to the organization and project team.

Dominant Culture

The establishment of a dominant cultural model is the most critical step of the three. Doing this involves formal recognition that company culture is the dominant one. All other cultures are subordinate. Since a company's culture is strongly influenced by the national culture where the company is headquartered or maintains its largest base of operations, the cultural norms of that nation will be dominant as well. Therefore, the way a multinational organization and its project teams operate is significantly influenced and driven by ideals and behaviors of the company. This is why there can never really be a merger of equals. Rather, one company has to remain dominant in order to establish a base set of operating principles. Those base operating principles guide the way the project teams conduct their work and interact with one another.

Integration of Team Identity

In order for multinational project teams to function effectively, however, their members must share a common set of goals, values, motives, and operating norms that they can all relate to and follow. We refer to this as integration of team identity. Within the company culture, the project team must establish its own project culture. The project culture is still dominated by the way the company does business, but it is modified by the cultural values and behaviors represented by nations participating on the project team.

As described earlier in this chapter and in Chapter 2, the team charter is a critical project artifact for any team, but especially so for a multinational, multicultural virtual team. Within the team charter, the team norms define how the team will interact and incorporate the cultural nuances of the team.

Continuous Cultural Awareness

The final step in the convergence of multiple cultures is continuous cultural awareness. Virtual project managers must be the champion for cultural awareness because at the project level, cultural awareness goes beyond just being aware of cultural differences between team members. Cultural awareness at the project level has everything to do with trust and relationships. As we discussed in Chapter 6, the project network on which a virtual team is built is held together by

relationships, and trust is the foundation of those relationships. The cultural dichotomies discussed earlier and the variations in behavior that result from those dichotomies provide the opportunity for mistrust to arise when team members are culturally ignorant. For example, some team members may feel that addressing a problem or a source of conflict affecting the team is confrontational, but it is necessary to avoid distrust and fragmented relationships. Being culturally aware of the presence of these cultural issues allows project managers to work directly with affected individuals.

Different cultures rely on different mechanisms for deciding if other members on the team are trustworthy.[36] People from some cultures assume everyone has good intentions until they prove otherwise. Others see trust as non-existent until earned. Some people gauge trust on other people's performance, while people from other cultures look for endorsement from a trusted third party to establish trustworthiness.

All these cultural differences are present in multinational enterprises and organizations. The convergence of the differences will occur where the "feet hit the street," as the saying goes—at the project level. Cultural differences, of course, add more complexity to already complex virtual project environments. Much is left to virtual project managers to work and resolve while managing the project and leading the team. However, some significant organizational factors must be in place to assist an organization's virtual project managers. These organizational factors are discussed in the next two chapters.

Assessing Cross-Cultural Awareness

The Cross-Cultural Awareness Assessment measures the cultural awareness capabilities of a project team. The evaluators, organizational management, the project sponsor, and the virtual project team leader are responsible for evaluating and assessing the cultural competence on a project. Essentially, they are looking for a proper and balanced set of anticipated cultural skills that will raise the probability of success for team formation and interaction over the life of the project.

It is recommended that the members noted above complete the assessment. Once completed, meet to discuss results and findings as well as detail next steps (if warranted) to increase cross-cultural awareness.

Cross-Cultural Awareness Assessment

Date of Assessment: _____

Virtual Project Team Code Name: _____

Virtual Project Manager Name: _____

Project Start Date: _____

Planned Completion Date: _____

Team Size: _____ Number of Countries Represented: _____

Assessment Completed by: _____

Confidential Assessment: _____ Yes, confidential

_____ No, not confidential

Assessment Item	Yes or No	Notes for All No Responses
Project team members have participated in formal cultural awareness training.		
The project consists of a high percentage of team members with prior experience working on cross-cultural project teams.		
Project team members have country- and culture-specific knowledge and experience specific to the countries represented on the project.		
Project team members have proficient (written and verbal) language skills in the primary language of the project.		
Project team members have effective communication (especially listening) skills.		
A cultural assessment has been performed for the project.		
Cultural factors with wide variation within the team have been identified.		
Cultural-based strategies have been developed for the project.		
Project team members have confidence working with team members from other cultures.		

Assessment Item	Yes or No	Notes for All No Responses
Project team members have practical knowledge and ability to work successfully with power distance differences.		
Project team members have practical knowledge and ability to work successfully with uncertainty avoidance differences.		
Project team members are knowledgeable about and do not exhibit gender bias.		
Project team members have practical knowledge and ability to work successfully regarding the cultural dimension of restraint.		
Project team members have practical knowledge and ability to communicate in both low and high context settings.		
The project team comprises both divergent and convergent thinkers.		
A healthy level of assertiveness has been displayed in cross-team communication.		
The project manager plans to visit all distributed work sites at least once per quarter.		

Findings, Key Thoughts, and Recommendations

Notes

1. Pernille S. Strøbæk and Joachim Vogt, "Cultural Synergy and Organizational Change: From Crisis to Innovation," *Journal of Business and Media Psychology* (August 2014). http://journal-bmp .de/2013/06/english-cultural-synergy-and-organizational-change-from-crisis-to-innovation/?lang=en.

2. Bill Richardson, "Culture-Induced Complexity: What Every Project and Program Manager Needs to Know," PMI White Paper (Newtown Square, PA: Project Management Institute, Global Operations Center, 2014).

3. Robert Mitchell, "Cultural Intelligence: Everybody Needs It," *Harvard Gazette* (November 3, 2014). http://news.harvard.edu/gazette/story/2014/11/ cultural-intelligence-everybody-needs-it/.

4. Richardson, "Culture-Induced Complexity."

5. Parviz F. Rad and Ginger Levin, *Achieving Project Management Success Using Virtual Teams* (Plantation, FL: J. Ross, 2003).

6. Terri R. Kurtzberg, *Virtual Teams: Mastering Communication and Collaboration in the Digital Age* (Santa Barbara, CA: Praeger, 2014).

7. Richardson, "Culture-Induced Complexity."

8. Michelle LeBaron, "Culture and Conflict." In G. Burgess and H. Burgess, eds., *Beyond Intractability* (Boulder, CO: Conflict Information Consortium, University of Colorado, July 2003). http://www.beyondintractability.org/essay/ culture-conflict.

9. Frank T. Anbari et al., "Managing Cross Cultural Differences in Projects." In PMI Global Congress, Orlando, FL (Newtown Square, PA: Project Management Institute, 2009).

10. https://geert-hofstede.com/national-culture .html, accessed June 2016.

11. Geert Hofstede, "Dimensionalizing Cultures: The Hofstede Model in Context," *Online Readings in Psychology and Culture* 2, no. 1 (2011). 10.9707/2307-0919.1014.

12. John W. Bing, "Hofstede's Consequences: The Impact of His Work on Consulting and Business Practices," *Academy of Management Executives* 18, no. 1 (February 2004). http://www.itapintl.com/index.php/about-us/articles/hofstedes-consequences.

13. Hofstede, "Dimensionalizing Cultures."

14. Margaret R. Lee, *Leading Virtual Project Teams: Adapting Leadership Theories and Communications Techniques to 21st Century Organizations* (Boca Raton, FL: CRC Press, 2014).

15. Bing, "Hofstede's Consequences."

16. Hofstede, "Dimensionalizing Cultures."

17. Deborah L. Duarte and Nancy T. Snyder, *Mastering Virtual Teams: Strategies, Tools and Techniques that Succeed* (San Francisco, CA: Jossey-Bass, 1999).

18. Bing, "Hofstede's Consequences."

19. Anbari et al., "Managing Cross Cultural Differences in Projects."

20. Edward Twitchell Hall, *Beyond Culture* (New York, NY: Anchor Books, 1976).

21. Anbari et al., "Managing Cross Cultural Differences in Projects."

22. Lee, *Leading Virtual Project Teams*.

23. Duarte and Snyder, *Mastering Virtual Teams*.

24. Hofstede, "Dimensionalizing Cultures."

25. Anbari et al., "Managing Cross Cultural Differences in Projects."

26. Cecilia Cheng and Woo Young Chun, "Cultural Differences and Similarities in Request Rejection: A Situational Approach," *Journal of Cross-Cultural Psychology* 39, no. 6 (November 2008): 745–764. doi: 10.1177/0022022108323808.

27. Robert J. House, "How Cultural Factors Affect Leadership," July 1999, Wharton School of the University of Pennsylvania. http://knowledge.wharton.upenn.edu/article/how-cultural-factors-affect-leadership/.

28. Kurtzberg, *Virtual Teams*.

29. Stephanie Quappe and Giovanna Cantatore, "What Is Cultural Awareness, Anyway? How Do I Build It?" http://www.culturosity.com/pdfs/What%20is%20Cultural%20Awareness.pdf.

30. Duarte and Snyder, *Mastering Virtual Teams*.

31. Lee, *Leading Virtual Project Teams*, p. 117.

32. Lothar Katz, "Organizational versus National Culture," 2005. http://www.leadershipcrossroads.com/mat/Organizational%20vs%20National%20Culture.pdf.

33. Interview with Mary Dunkin. Conducted May 2016.

34. Katz, "Organizational versus National Culture."

35. "Is Your Company Culture Too Strong?" *Kellogg Insight* (July 6, 2015). www.insight.kellogg.northwestern.edu/article/is-your-company-culture-too-strong.

36. Kurtzberg, *Virtual Teams*.

8

USING TECHNOLOGY TO COMMUNICATE AND COLLABORATE

Gone are the days when a project's success is measured only by the accomplishment of the triple constraints. We now know that this is an overly simplistic and narrow view of success, largely because it ignores the human element.[1] We now correctly view projects as a set of technical challenges that must be overcome by a group of people who must work collaboratively and collectively toward the achievement of a common goal. Doing this requires that the set of people be able to share their ideas; coordinate their activities; discuss their insights, perspectives, and opinions; and collectively solve problems. It is this high level of human interaction that, frankly, makes project work so challenging. For virtual project teams, this challenge is compounded by the fact that team members are separated by time and distance, making human interaction even more problematic. As pointed out in Chapter 1, one of the primary differences between traditional projects and virtual projects is that the virtual project team must communicate and collaborate almost exclusively through the use of technology.

Although technology is used for communication and collaboration on traditional projects as well, the co-located nature of traditional projects allows for a high degree of direct human interaction between team members. For virtual teams, all interaction has to be facilitated through the use of tools. This is not necessarily bad, however, because technology is the single most important enabler that allows people around the world to cross physical, organizational,

and cultural boundaries and to work as a team, albeit sometimes a widely distributed team.

It is as true today as throughout history, projects and the people who make up those projects are expected to produce results. Today, however, time, distance, and communication barriers are in play to prevent virtual project teams from working effectively. If used correctly, technology is the means to overcome these barriers.[2] Additionally, with the introduction of social media tools on projects, it is now possible for team members who are separated by hundreds or thousands of miles to establish a base level of personal relationships. As the effective use of technology increases, so too can the overall performance of the virtual team and the results produced. The key word in this truism is *effective*.

Even though technology has the potential to enable communication and collaboration, it is also often the source of great frustration, conflict, and hindrance to effective collaboration on a virtual project if not planned and used purposefully. Project managers must assume the responsibility for ensuring that technology is indeed used effectively and serves as a communication and collaboration enabler on their projects. Doing this requires the establishment of project-level technology usage requirements and tools selection based on the usage requirements. Getting usage requirements right begins with an understanding of the role of technology on virtual projects.

Role of Technology

The primary role of technology in distributed teamwork is one of helping the project team overcome the challenges created by time and distance separation.[3] Successful use of technology on a virtual project hinges on four points:

1. Understanding how the team will communicate and collaborate.

2. Understanding how technology can be leveraged to improve that communication and collaboration.

3. Matching technology selection to communication and collaboration methods and practices.

4. Using the technology efficiently to improve the team's performance.

Many times large investments in project technologies have failed to pay for themselves because the technology was selected based on false financial estimates of productivity gains or because of pressures to implement technology solutions used by industry leaders rather than selecting solutions that integrate into the team culture and needs. In nearly all successful examples of technologically enhanced team communication and collaboration, the technology was selected based on how well it enhanced established communication and collaboration methods and processes.[4]

Project managers should select electronic technologies that best meet the needs and usage of their virtual teams and that integrate with the current suite of tools used within the organization. Best practice is to use technology tools that complement the culture and that are selected based on need. Technology should always increase team productivity and should be evaluated based on that measure.

There is no ideal set of technologies for all teams. There are basic technologies, however, that most all teams will benefit from—such as the telephone, email, and calendaring and scheduling systems—but since all organizations are unique, so too are their virtual communication and collaboration needs. Project managers therefore must develop a clear strategy for matching technology options to the communication and collaboration needs of their virtual project teams.

Think Usage First

It is quite common for project managers to adopt the technology tools most widely used within their organizations without much forethought about *how* technology should be used to facilitate interactions among their virtual team members. Also common is for a firm's information technology (IT) department to select, rollout, and mandate the use of certain technology tools without first consulting project teams about their business processes and methods. What makes this worse is that project managers know that enterprise-level tools can be ineffective on project teams. This knowledge is consistent with findings from virtual team studies, which show that the vast majority of virtual teams rely on email, telephone calls, and occasional video conference meetings to do their work.[5] It is important to note that technology will not solve communication problems, but it can help to improve them. Always remember that the primary focus is not on the tools themselves, but on *using* the tools to facilitate a culture of knowledge sharing, information sharing, relationship building, decision making, and working collaboratively.[6]

Using Technology to Communicate

Communication is *the* most critical factor in determining success on any project, virtual or traditional. This is because project success hinges on the sharing of information among team members. The greatest barrier to effective communication on a virtual project is, of course, the physical distance between team members. This physical distance, coupled with the resulting time zone differences, creates an obstacle to the most effective method for information sharing: person-to-person conversation and interaction. Project communication on virtual projects must therefore move to technology platforms.

Four primary usage factors must be considered when selecting communication tools for virtual projects:

1. Communication type
2. Time
3. Location and forum
4. Inclusiveness

Communication Type

There are three primary ways in which members of a virtual project team interact: through conversations, through transactions, and through collaborations.

Conversational interaction is a free exchange of information among team members for the primary purpose of knowledge exchange, discovery, or relationship building. Conversational interaction is free of constraints and does not use a central repository of information.

Transactional interactions occur when a team member exchanges something tangible (e.g., a requirements document, project plan, or design specification) with one or more members of the team. The transaction usually is constrained by a need and expectation (stated or implied) of the receiver of the deliverable.

Collaborative interaction occurs when two or more members of a team work together to complete a task, solve a problem, brainstorm new ideas, or develop a common deliverable. Collaborative interaction requires a common electronic workspace and a repository to store completed and in-process work.

Time

Once virtual project team members understand the types of communication interaction they will be engaged in, the next step is for them to understand which method of communication best meets their needs from a time perspective—synchronous, asynchronous, or a combination of both.

Synchronous communication is direct communication among team members, where the communications are time specific.[7] This means members are present and communicating at the same time, but may be separated by geography and time zones. Synchronous communication methods are best for interactive activities, such as brainstorming, problem solving, decision making, and team status reporting.

Asynchronous communication occurs when team members involved in the communication are not present at the same time. Therefore, time delays occur between communication exchanges between team members. Asynchronous communication methods work well for information and data exchange that does not have to be completed in real time.

Location and Forum

The next primary technology usage factor that needs to be considered is the location of the virtual project team members. This is influenced by the type of virtual project, as described in Chapter 3.

Mostly Co-located Virtual Projects

Communication on a mostly co-located virtual project is dominated by direct communication between the co-located members, but also relies heavily on communication with all distributed team members. When choosing the technology tools for this project type, project managers must be very aware of what information can be fully exchanged between team members without the use of technology and what information has to be exchanged via technology to ensure the inclusion of the distributed team members. For mostly co-located teams, communication forums can span from face-to-face to 100% virtual exchanges. Project team members who are co-located can exchange project information through face-to-face meetings or at least through telecommunication means. In this type of project, it is not uncommon for co-located team members to assemble in a single conference room, with distributed members joining them through synchronous audio or video conferencing technologies.

Nationally Distributed Virtual Projects

Communication on a nationally distributed project is accomplished exclusively through technology. Since team members are separated by a few time zones at most, a blend of synchronous and asynchronous communication is possible. Team members on a nationally distributed virtual team adhere to a single dominant national culture and language (although many project managers claim that working with team members from different regions of a country is sometimes like working with completely different cultures). Because of the national commonality, additional technology requirements for cultural and language nuances are not required. Most communication can be performed by audio or video conferencing as well as simple electronic exchange through email or collaboration sites.

Internationally Distributed Virtual Projects

Team members on a global project can be separated by great time and distance, causing synchronous communication to be impractical. Since nearly all communication is performed virtually, technologies that support centralized and asynchronous communication and information sharing are required. Additionally, since the dominant language used to communicate (both verbally and in written form) is likely to be a second, third, or fourth language to many or most team members, additional technology usage requirements, such as language translation tools, may be needed to support language differences.

Inclusiveness

Inclusiveness simply refers to how far the sharing of information will permeate across the team and within the organization. Figure 8.1 illustrates the four primary communication populations for a typical project, virtual or traditional.

Need to Know

On some projects, information may need to be protected, and communications pertaining to that

Figure 8.1: Primary Project Communication Populations

information have to be restricted to project team members with a need-to-know classification. This is common on some government projects and highly sensitive strategic projects. In such cases, technology that can enable the need-to-know information restriction requirements for a particular project is required.

All Project Team Members

Other projects limit the sharing of information to all project team members and exclude the communication and sharing of project information outside of the project team. This is an extension of the need-to-know usage and also requires technology that can enable restricted access to project team members only.

Project Stakeholders

Communication and information sharing outside of the project team may be limited to a specific set of individuals who have a stake in the performance and outcome of the project. Communication outside of the project team will be limited to these identified individuals.

General Organizational Population

For most projects, communication and sharing of project information is unrestricted within the confines of the organization within which the project operates. In such cases, project team members can utilize generally available means of electronic communication used within the company.

Using Technology to Collaborate

In a business model where the work of a project team can be digitized, disaggregated, distributed across the globe, modified, and reintegrated into a holistic solution, a high level of collaboration between distributed team members is required. *Collaboration* generally refers to individuals working together to create ideas, to solve problems, and to deliver outcomes.[8] On projects, *delivering outcomes* usually refers to completing tasks and generating deliverables. Project teams perform their work in a combination of individual efforts and collaborative interactions with other members of the team. Interactive collaboration is when two or more people on the project work together through idea sharing and thinking to complete a task, solve a problem, brainstorm new ideas, or develop a deliverable. It is teamwork taken to a higher level. With the changes and advancements in technology, such as high-speed internet, web-based programs, file sharing, email, and video-conferencing, collaboration has become a more productive way of completing project tasks. Interactive collaboration on virtual teams, however, is more difficult than on traditional teams due to the inherent differences in time, distance, and culture. Virtual collaboration, therefore, must be facilitated through technology.

As with the selection of communication tools, how the team collaborates must be considered *before* the particular technologies to be used on a virtual project are selected. Three primary usage factors must be considered when selecting collaboration tools for virtual projects:

1. Team tasks
2. Project workflow
3. Workspace

Team Tasks

Understanding collaborative interaction on a virtual team begins with looking at the relationship between the types of tasks undertaken by the virtual project team and the technologies they use to help them complete those tasks. We generalize team tasks into two primary categories: low-complexity tasks and high-complexity ones.

Low-complexity tasks are those that involve minimal interdependence on other project tasks and require limited collaboration between team members. Most conversational and transactional interactions fall into this category. For these types of tasks, asynchronous technologies, such as email, blogs, data repositories, and websites, are completely sufficient.

High-complexity tasks are those that involve a high degree of collaboration and information sharing among team members in order to complete a task. Team members are very interdependent when working on high-complexity tasks, and the technologies they use must support both synchronous and asynchronous collaboration of interdependent workflow. For these tasks, technologies such as videoconferencing, electronic whiteboards, common workspaces, and shared data repositories are most beneficial.

Project Workflow

Workflow is the sequence of steps or the process through which a piece of project work passes from initiation to completion. Simply, it is how project tasks get completed through the interaction of team members. Generally, project workflow is made up of two types of activities: sequential and parallel (sometimes referred to as concurrent).

Sequential tasks imply a finish-to-start dependency among team members. One team member cannot begin his or her task until another task is completed. Parallel tasks are those that can be performed at the same time, either independently or interdependently. Parallel tasks may have an additional requirement that task outcomes be integrated with one another once they are completed.

Both types of tasks require that all team members are on a common electronic workspace that provides adequate visibility to the progression of work in process or completed on the project.

Workspace

When it comes to collaborative technology, the third usage factor is workspace. There is a big difference in technology selection if workspace is shared versus private, desktop versus laptop, and open space versus closed office. Some technologies for collaboration work well in some settings, but not in others, and none of the technologies available today work well in all settings. One size does *not* fit all. Like the selection of other technologies, workspace collaboration tools should take into consideration the organizational culture, the user, and project needs.

Technology Options for the Virtual Project

In the last few decades of the 20th century, the invention and introduction of communication and collaboration technologies was a driving force behind the rapid expansion of virtual projects as a means to execute an organization's work. Since that time the number of new technological tools has exploded; there are so many, in fact, that choosing the tools needed has become complicated. This overabundance of choices often results in technology overload, which causes project managers to defer selection to the firm's IT department.

There is large risk in this approach, however. IT departments are, of course, made up of technologists who see the world through the technological lens. Left unchecked, their choice of tools can be primarily influenced by the newness of a technology or the number of features available, leaving the use of the technology, the usability, and the robustness and reliability as secondary considerations.

In this chapter, we are recommending an alternative approach, one that considers technology usage as the primary factor. After first considering how a virtual project team communicates and collaborates via technology, the next step in the selection process is understanding the types of technologies available to a virtual team. (See Figure 8.2.)

The sections that follow provide categories of technologies that support communication, collaboration, and relationship-building interactions. Since specific tools tend to come and go over time and enterprise tools from dominant commercial suppliers try to be all things to all users, it is highly recommended that technology *categories* be evaluated before specific tools are considered.

Messaging Tools

Messaging systems are the most ubiquitous of virtual team interaction tools. They enable direct communication among virtual project team members both asynchronously and synchronously. Further, depending on the tools selected, messaging tools scale from direct one-to-one communications to many-to-many large forum communications. What follows is a discussion of the most widely used messaging systems at the present time.

Email

Email is now the primary means of communication on a virtual project, replacing the telephone as

Figure 8.2: Virtual Project Technology Selection Process

the tool of choice. It is also the simplest of tools to use and involves one person (the email originator) composing an electronic message that is sent directly to one or more recipients. When time and distance prevent physical one-to-one conversations and direct information sharing among virtual team members, email keeps virtual projects moving forward. The advent and proliferation of wireless technologies enable email to be accessed from nearly anywhere on the planet.

Email is used to provide status communication, make inquiries, share and review documents, and coordinate logistically.[9] It is an asynchronous tool that provides a recipient the option of when or if to reply. For this reason, email is not necessarily a good choice for communication when a rapid response is required. Also, email is not a good choice for communicating controversial messages or for resolving conflicts among team members unless other, richer forms of communication technologies, such as direct phone or video conferencing, are unavailable. Without the added richness of voice, eye contact, and body language, emails are easy to misinterpret.[10]

Under normal circumstances, it is not uncommon for virtual project team members to receive dozens of emails daily, creating a response backlog and delay. To help manage this scenario, virtual project managers and other team members must become proficient at crafting effective business emails. For some tips, see the box titled "Effective Email Practices for Virtual Project Teams."

Effective Email Practices for Virtual Project Teams

As our professional communities become more global, we find we often work with people who are in far-flung places, at least far from where we are. This means we will be communicating with teammates who don't share the same first language as us or the same cultural norms as us. We therefore need to develop more effective global communication skills so that we consistently deliver business results—even when the rest of our team or organization is scattered across the globe. Collaboration using email is one way virtual business is conducted, but it comes with a number of limitations, which you may have encountered:

- Many times it is unclear what action you are expected to take after reading an email.

- Emails received many times contain misspellings, missing words, or grammatical errors that obscured the sender's meaning.

- Sometimes, it is difficult to know if an email message you send will be interpreted as intended by someone whose language and culture are very different from your own.

- Many times, it takes several emails in order to clarify an issue with your peers in other countries for whom English is a second language.

Email is convenient. We can store it, delete it, and react to it instantly. Yet it is the immediacy of email that leads to the problems just listed if an email is sent without thinking deeply about what has been written and without checking for errors, expectations, outcomes, and completeness.

Above all, it must be recognized that words alone constitute only about 10 percent of overall communication; the rest is made up of facial expressions, body language, tone, and other physical manifestations. Therefore, email starts with a handicap as it only contains words—users are, in effect, sending only 10% of the communication.

What to Do

Jose Campos, founder of Rapid Innovation, has provided these suggestions for improving the quality and consistency of business email in a distributed environment:

- *Write and send with care.* Email is a very casual mode of communication, and we often choose to use it to discuss critical issues. When we do, it is vital to read and write with the importance and focus that the issue deserves. Focus your attention to ensure that you give the message your full concentration.

- *Take full advantage of the subject line.* The subject line should give a clear idea of content and action needed. When sending a reply, change the subject line if it needs changing; do not just drag out an old email from weeks ago and reuse it to cover a new topic without changing what the subject line says. And, whatever you do, do not send the message in the subject line, as the limited characters available do not provide adequate context for the reader.

- *Cover one item of business in each email message.* Emails have replaced the printed memo as announcements, reminders, and calls to action. Use separate messages with individual subject headings to make it easier for readers, allowing them to understand, review, and react to each subject logically.

- *Clarify what actions are required of each recipient.* Ambiguity of actions required can cost the virtual team precious time. Remember that an action item is made up of three parts: (1) do what, (2) by whom, and (3) by when. For example, "John, please develop a database of all midwestern customers by the end of this week. I will need the database by Friday, Nov. 16, at 2 PM EST." Anything less than that and it is not an action item.

- *Be wary of emotionally charged content.* If you have something emotional to share, it is best to do it in person or on the phone. Understanding is gained through voice, body language, and facial expressions. Too many people use email as a shield to avoid unpleasant confrontations, and they end up alienating the recipient. If you would not say something to someone in person, do not say it in an email.

- *Do not use email to deliver the really bad news.* Never use email to chastise or criticize, and never (never!) to deliver any seriously bad news. A message that may seem reasonable to you can hit another person hard, and you will not know it since you cannot see the reader's non-verbal cues. Any message that will have a major effect on an employee's work or life should be delivered face-to-face or at least over the phone.

- *Be professional.* Minimize the use of underlines, boldface, all caps, and exclamation points. These all have the same unpleasant effect as shouting at someone face-to-face.

- *Acknowledge receipt and respond promptly.* In many cases, the sender is waiting for a reply from you, even if it is just an acknowledgment that you received the message. Consider that things may be "on hold" until you reply, which can impact productivity and decision making. When you receive a request or an action item, take immediate action. For example, if someone is asking for a meeting date, let him or her know immediately of your availability or alternatives.

- *Be judicious in your use of group addresses.* There may be no need to copy all members of the team on every email. Doing so decreases teammates' productivity and often leads people to assume that someone else will handle the issue. The hope is that someone else among those people will do it, so "I don't have to."

- *Read and edit before you send.* As a form of written communication, email represents you professionally, so send your best. And for particularly important messages or any message composed in anger, treat your message as if it were a traditional letter: write, print, read, revise, and then send. And remember, never send an email without first checking the spelling. Typos and other blemishes make the message difficult to read and lower its professional quality.

- *Do not send or read email at inappropriate times.* Email has become an addiction, but the constant pursuit of an email fix may be costly. One of the newer forms of poor office etiquette is paying more attention to email than to a conversation or business meeting. It has been proven that if people are attending to multiple things at the same time, they are not going to retain as much information and will make more errors than if they were focusing on one thing. This can lead to lost productivity and wasted dollars because people are not focused on their work. It is best practice to dedicate periods of time during the day to read and respond to email as a single task.

Let's face it. Email is a convenient, fast, and valuable tool for the distributed team and virtual business environment. Like any tool, it can facilitate or hinder communication. Used effectively and in concert with face-to-face meetings, email is an exceptional way to share information, foster creative thinking, and facilitate critical personal meetings.

No virtual team should rely exclusively on email (or any other form of one-way communication) to share information and maintain business relationships. Recent advances in email enable vocal and visual enhancements to the traditional text-based option. Audio and video email allow senders to incorporate a voice message or short video message to traditional email messages. These capabilities are currently add-on features, but look for them to be integrated into enterprise-level messaging systems in the near future.

Instant Messaging

A technology that originated on private personal computing platforms, instant messaging (IM) has moved into the business environment in a big way. It's primary value to the virtual project team is its use as an "electronic water cooler" where team members can engage in quick, synchronous conversations relating to both personal and work topics. The proliferation of IM in the business environment

has led to a decrease in the use of chat rooms due to the increased privacy and synchronous communication ability that IM provides over chat rooms. However, because IM occurs in real time, team members must be online at the same time to realize its true value.

With the advancement of wireless and mobile technologies, IM participants now have the ability to be away from their normal workstations and place of work to participate. This further frees virtual project team members from the constraint of geographical presence.[11]

IM is best used for quick information exchanges, queries, clarifications, advance notifications of other communications (such as an important email), project logistics (meeting reminder or time change notification), and social interaction. Participants should refrain from using IM for task-based communication or project-related decisions due to the lack of recording and reporting capabilities. Listed in the following box, "Dos and Don'ts for Instant Messaging in the Workplace."

Dos and Don'ts for Instant Messaging in the Workplace

As with all communication tools used in the workplace, IM can be used appropriately or inappropriately. We have witnessed its use in both ways and offer a few guides.

Instant Messaging Dos

- Do use IM for informal conversations on both work-related and personal topics that are appropriate. IM is a great way to create a water cooler feel for chatting online. It can be used for one-on-one or group discussions and provides an excellent informal mode of communication and collaboration for virtual project teams.

- Do use IM to help build personal connections to other virtual team members. IM provides users the opportunity to check on the physical and emotional states of other team members that other group-focused team interchanges don't necessarily allow for.

- Do use IM as a means for improving team efficiency and project productivity. Key information exchanges can be taken off-line between other synchronous team meetings to speed up communication and workflow. Likewise, IM can be useful for addressing project-related issues and problems that need immediate attention.

- Do use IM for situations where virtual team members may have language challenges. Verbal communication may be frustrating and confusing to some, and written communication via IM provides an excellent alternative. IM can also be used to solicit input from team members who are quiet during larger verbal team exchanges.

- Do use IM to establish a direct link between the project team leader and remote team members. Doing so establishes a direct line to the team leader that is critical for effective communication.

Instant Messaging Don'ts

- Don't use IM for communications that may include sensitive or confidential information. IM is subject to potential hacking, viruses, phishing scams, and other computer-based problems that may jeopardize sensitive information.

- Don't use IM beyond appropriate casual and informal project team discussions due to the potential for legal issues and risks. Remember that all IM activity creates a historical record of communications that can be accessed.

- Don't use IM for complex issues or problems that have a high potential for creating misunderstanding. Difficult issues are best addressed via means where full context can be effectively communicated.

- Don't use IM for communications where the potential is high for significant conflict or where past interactions have damaged trust. It is better to seek other communication options, such as face-to-face meetings, if possible.

- Don't use IM where you sense or know someone is experiencing serious emotional difficulties. IM as a communication vehicle does not work well in situations where one needs to provide special attention to another person's emotional well-being.

List Server Applications

A list server is simply an automated email distribution list that is used to broadcast communications to a group of people, such as a virtual project team. The automation feature is primarily the ability for people to subscribe and unsubscribe to a particular list server.

List server tools can be used in several different ways on virtual projects. The two most common uses are broadcast email communications and discussion groups. Broadcasting common email messages to project team members on a distribution list is wonderful for communicating such things as schedule changes, requirements changes, status reports, meeting notes, decision outcomes, and so on. Some teams create project newsletters that are distributed via list server applications.

Discussion groups can be used to facilitate interaction among members of the virtual team. Normally, discussions are focused on a single topic, such as gathering functional requirements or generating ideas to solve a particular technical problem. Additionally, discussion topics can be informal in nature and used to build and foster cross-team relationships. Discussions on favorite hobbies, sports teams, or holiday experiences are great ways to learn more about virtual team members beyond what capabilities they bring to the project team.

List servers also provide threading features that enable historical tracking of discussions. Threading allows team members to familiarize themselves on a topic by sorting on date, sender, and keywords. However, broadcast emails via list servers also have limitations. Since they are a form of asynchronous communication, the likely result is a lack of immediate response that can cause time delays. Also, broadcast emails, like all email, likely lack needed context, which cause some confusion and possible misunderstanding for some distributed team members.

Weblogs

Weblogs (more commonly referred to as blogs) are web-based versions of broadcast email and list servers. The differences are the medium used and the fact that blogs are individually maintained journaling sites.[12]

Weblogs are a technology that originated in the public domain. Blogs have been fully integrated into the corporate world. Similar to list servers, blogs facilitate communication among team members and track team communication and decision making, and can be used as communication vehicles among the project team and stakeholders, clients, and other business partners. Blogs can serve as the means for both internal and external project communication.

Blogs introduced special features, such as "like" capabilities and "comment" fields, that facilitate rapid and direct communication between an author and his or her recipients.

Electronic Bulletin Boards

On virtual project teams, the primary use of electronic bulletin boards is to share project knowledge and information. Electronic bulletin boards are wonderful tools for closing the distance gap among virtual team members and for aligning disconnected team members to the project. Electronic bulletin boards are used in the same manner as physical bulletin boards found on the walls of public places: to post important and interesting information and communications.

Important baseline project information, such as project goals and metrics, the project charter, the business case, requirements documentation, and master schedule, can all be posted to an electronic bulletin board for quick access by all team members. Additionally, current information on project status, decisions, and changes are also important content for team members to be able to access in a central location without having to search a document repository. Electronic bulletin boards can be vital tools for orienting new team members to a project.

Electronic bulletin boards can also be used to facilitate extended group discussions among virtual team members. They provide all the same threaded discussions as list servers, but have the added benefit of access through a company's intranet or through the public internet.

Conferencing Technologies

Conferencing tools were introduced into the business environment in the early 1980s and are now commonly used in every industry. They enable richer real-time interaction among virtual team members by providing both verbal and visual communication. Expressions of understanding, confusion, satisfaction, concern, and so on can be easily seen and immediately addressed, informing the team more quickly and effectively than a trail of emails, voicemails, and IMs. For this reason, conferencing tools facilitate quicker contextual understanding among team members and therefore a deeper level of communication and collaboration interactions among virtual project team members.

Telephone Conferencing

Although the telephone and Voice over IP (VoIP) are used less than in previous decades, they are still invaluable tools for one-to-one or larger team discussions. The telephone is the preferred tool for highly interactive, immediate, and fast-moving communication. Voice and verbal cues provide a much richer communication than text-based options and are performed synchronously.[13]

However, a major drawback of phone conversations is the lack of documentation of the communication that took place. For virtual project teams, which can be distributed across the globe, communication documentation *is a must*. Therefore, all important virtual project team telephone conversations must be followed up with text-based documentation that chronicles the major points discussed, actions to be taken, and decisions that were made. The text-based report should then be distributed to team members and stakeholders who have a vested interest in the conversation. If possible, the phone conference should be recorded and made available to those with a need to know and other interested parties, if the conversation is not proprietary or confidential.

Video Conferencing

Although similar in use to audio conferencing systems, video conferencing tools provide a richer communication experience by including visual images of those communicating with one another. They provide some of the dimensions of face-to-face communications, such as the ability to see body language or visualize a common document while meeting. A side benefit to the use of video conferencing tools is that since conference participants can be seen by others, behind-the-scenes multitasking is greatly reduced or eliminated. This makes it easier for project managers to hold participants to meeting etiquette and ground rules, as discussed in Chapter 6.

Video conferencing tools are now effective in computer-based, mobile platform-based, and room-based formats. Today's computer-based video conferencing systems are viable options for virtual project teams that are highly distributed in many geographical locations. For effective computer-based video conferencing, all that is needed is a personal computer with a camera, VoIP, and video conferencing software capabilities. The internet connection and bandwidth can be limiting factors in some locations, causing temporary loss of audio and visual communication. Such interruptions are mildly to severely distracting and disruptive to conference participants, so we recommend always having an audio-only backup.

Advances in internet connectivity and bandwidth that supports mobile platforms, such as phones and tablets, mobile-based video conferencing extends the limits of geographic distribution of virtual project team members. Mobile-based video conferencing makes it possible for project team members to communicate and collaborate anywhere on the planet where an internet connection is available. Mobile-based video conferencing can greatly reduce project cycle time by eliminating lost, slow, or non-existent communication among team members who reside and work in different locations.

Many organizations now have room-based video conferencing systems located at their major sites. This is a great option for virtual project teams that are distributed but have co-located members at several sites. Room-based video conferencing tools allow for a good combination of both face-to-face and remote video conferencing communication and collaboration. Most room-based video conferencing systems also support connection to desktop video conferencing systems, so remote project team members who are not part of a co-located site can participate as well.

Although more costly to purchase and install than desktop or mobile video conferencing tools, room-based video conferencing provides the richest virtual communication and collaboration experience. The integrated solution of video monitors, cameras, audio, and video conferencing software is the simplest and highest-quality solution as long as the broadband internet connection is sufficient in all locations.

Webcasting

Webcasting tools are similar in function to video conferencing systems, but they have built-in limitations to simultaneous interaction among project team members. As a result, webcasting tools are more commonly used for knowledge exchange or information transfer sessions, such as senior management communications or group training sessions. Integrated document sharing and electronic whiteboard capabilities enable both audio and visual communication. Additional information can be provided via visual sharing that will serve to increase contextual understanding and overall communication effectiveness.

Most webcasts require a single facilitator who is in charge of driving the interaction between the main presenter and the webcast participants. Chat rooms or IM capabilities commonly are incorporated in webcasting systems, which become the primary means of input or feedback from participants to the facilitator or presenter. Additional features, such as polling, also enables group decision making that is effective in reducing time to decision.

Most webcasting tools have the ability to record a webcast session for later viewing and for meeting documentation, enabling asynchronous participation for team members who cannot attend a meeting in real time. The recording of sessions also provides a system of documenting meetings and can be used to certify critical project information exchanges and decision outcomes—such as a team decision to include a new product feature using the polling function of the webcasting tool.

The built-in two-way communication limitations mentioned earlier and the mix of audio/video interaction by the presenter and text-based interaction by participants make contextual misunderstandings more prevalent when webcasting tools are used instead of video conferencing tools. This unbalanced communication exchange requires a higher level of expertise on the part of the webcast facilitator to ensure overall understanding is maintained.

Collaboration Technologies

The power of a team, whether co-located or virtual, is that the collective wisdom of the whole is greater than the wisdom of any one member. Because of this, a team must have tools to support synergistic discovery, learning, and collaboration to tap the collective potential of all members. For virtual project teams, technology must go beyond just facilitation of real-time and asynchronous collaboration; it must also enable a participative project culture. The most effective tools for accomplishing this are chat rooms, team spaces, electronic whiteboards, groupware, and project management software.

Chat Rooms

Chat rooms are one of the oldest electronic technologies for supporting virtual project team collaboration, and they still have a place in business today. Chat rooms support both synchronous and asynchronous collaboration and can scale from peer-to-peer collaboration to large teams that need a common electronic forum to discuss project topics and solve project problems.

Since chat room participation can be restricted based on user credentials, chat rooms are excellent tools for specific conversations or for working with sensitive information. Bill Johnson, an IT project manager for an automotive parts manufacturer, uses chat rooms on a regular basis. According to Johnson:

> "I find chat rooms very useful for both team problem solving and brainstorming. Because most of our virtual team is co-located in a single site with a half dozen people working in other geographies, the remote members are more willing to bring up ideas in a chat room than on a conference call. I think this has a lot more to do with them being more confident with written communication than verbal. And frankly, their ideas are consistently some of the most creative on the team."

Some project teams use chat rooms for internal (cross-project) collaboration and external collaboration with clients and business partners. The simplicity of the technology allows for collaboration outside of a team's organizational boundaries. This makes chat rooms an excellent way to bring outside feedback, opinions, and input into discussions that normally would be closed to outside participants.

In addition, use of chat rooms is increasing as a method to provide instant feedback to customer inquiries. Customers often prefer to use live chat capabilities to avoid automated phone answering systems and lengthy email exchanges.

Team Spaces

A home base for team collaboration for traditional co-located project teams consists of the physical office of the organization within which the project team works. Collaboration occurs in conference rooms, project war rooms, cafeterias, and hallways of the office. For virtual project teams, there is no common physical office. Therefore, like all other things communication and collaboration related, the distributed project team must use technological tools to create a virtual home base. Such tools are commonly called team spaces.

The term *team space* is often applied to software solutions and web services that provide a toolkit for virtual teams to store and share files, discuss project matters, and collaboratively create project artifacts and deliverables.[14] These collaborative platforms become the center of team activities, and using them should bring greater efficiency, not extra, unnecessary work.[15] These spaces can also serve to keep team member profiles that enable individuals to learn about each other, their project roles, personal hobbies, and other shared information.

No two projects or project teams are alike; therefore, it should be expected that no two team spaces are identical either. However, standard guidelines for design and content should be established within an organization to facilitate navigation and use by various project stakeholders.

Team spaces have to be actively managed either by project managers or delegates working in their support. One of the more common team space management functions is access control. Various levels of access to particular team space pages, folders, and documents can and should be controlled based on the use and need to know of the various project stakeholders who will be accessing the team space. This is particularly true for stakeholders who are external to the organization.

Electronic Whiteboards

Electronic whiteboards serve the same purpose for virtual project teams as physical whiteboards in conference rooms do for co-located teams: They add the visual element to team collaboration activities. Whiteboards allow project team members to move beyond working individually on their desktop work stations to being able to visualize a work product and modify it in real time, collectively.

Whiteboards display images of documents, video files, drawings, and so on, which can be manipulated in a teleconferencing or video conferencing setting, saved, and later distributed. An example may be a team review of project expenditures using a complicated spreadsheet, with the expenditure items manipulated in real time by the presenter or other team members. Another common use is the creation of a project requirements document during a requirements-gathering work session among project team members and various project stakeholders.

The most effective electronic whiteboard tools support nearly any computer program or software application; interface with external devices, such as digital cameras, video cameras, or digital display projectors; and support interactive manipulation of whiteboard content from multiple computers.

Electronic whiteboards can boost the collaboration on a virtual project team in many ways. They can convert any standard meeting or presentation into a collaborative event. Electronic whiteboards allow team members participating in the meeting to be more engaged with the discussion. Files can be quickly accessed, edited, shared, and saved via the whiteboard, allowing team members to emphasize in real time any changes that need to be made or ideas that need to be considered. Electronic whiteboards enable the team to make persistent changes to any document or other project artifact. They also enhance communication for remote team members by enabling screen sharing so everyone in a meeting is focused on the same information.

Because electronic whiteboards can support nearly any stationary or mobile platform, they enable quick connectivity and a greater range of team interconnectivity and data sharing. This works great for today's mobile workforce. With the proliferation of touch technology, all information shared on an electronic whiteboard is now literally at your fingertips and can be manipulated through a human gesture.

Groupware

Groupware, which first emerged in the business environment about three decades ago, has experienced dramatic growth and widespread use in the past decade. The proliferation of virtual organizations and projects within those organizations that require a central location for team member collaboration has contributed to much of that growth.

Groupware, also referred to as collaborative software, creates an integrated, collaborative environment that enables virtual project team members, both co-located and geographically distributed, to collaborate with one another in local or remote locations via the internet or over an internal intranet. It is an evolving concept that supports multiple users working on related tasks, providing a mechanism that helps users coordinate and keep track of ongoing project assignments both synchronously and asynchronously.

By connecting project team members no matter where or when they are working, groupware serves to create a virtual office hub (see Figure 8.3) where all project information can be stored, shared, and worked on in a central location. This central virtual office hub makes finding information more effective than conventional means, such as emailing files and storing documents on team members' work stations.

Virtual project teams will find many uses for groupware applications including document creation and management, document and file exchange, document storage, project team calendaring and scheduling, task and deliverable management, threaded discussion, as well as chat and video conferencing. Like all integrated tools, groupware relies on ease of use (lower complexity = easier to use), willingness to use,

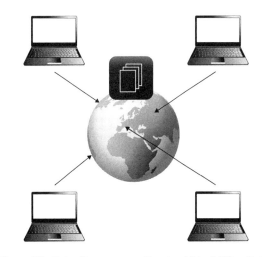

Figure 8.3: Using Groupware to Create a Virtual Office Hub

availability of the groupware for all team members, and an infrastructure that will support it in all locations where team members reside and work. Collaboration software works only if everyone on the team is able and willing to put it to use and realize its full benefit.

Project Management Software

It goes almost without saying that project management software is necessary for the successful execution of a virtual project. The good news for virtual project managers is that a suite of project management tools that is customized for virtual projects is not necessary. Rather, the most foundational and time-tested tools effectively translate from use on traditional projects to virtual projects. As we have stated throughout this book, however, *how* the tools are used will change in many instances.

Project management software is an essential tool to manage the workflow and tasking in the virtual environment. Primarily managed by project managers, the software is essentially used in the same way as on traditional projects. It is imperative, however, that everyone on the team is on the same software platform. Michael Washington, a mobile devices project manager, explains:

> "I recently led a virtual project that included a team overseas that we acquired several months earlier. Being the first project where the work of the two teams was highly integrated into a single platform, we soon came to the realization that we were unable to create an integrated project plan due to the fact that we used different project management tools—Basecamp and Microsoft Project. Because of institutional familiarity and initial investments made in the tools by each party, neither party wanted to convert to a new tool. The result was less than optimal and probably caused way more work on my part as the project manager than really necessary. For over half of the project, we managed tasking and deliverables through the use of a Microsoft Excel spreadsheet and workflow through the use of Visio. It took

months to reach a decision at the executive level on which common project management software solution would become the corporate standard."

Language Translation Software

The need for language translation software has increased significantly to assist communication between team members from various countries who speak different languages. The cause is certainly globalization.

Multinational virtual teams are normally cross-cultural teams with multiple languages represented, which magnify communication challenges. It helps when the virtual team can agree on a shared language to be used for the life of the virtual project. Then language translation software can be used as needed and appropriate for translation back into the team's shared language for work to be performed by other team members. There are many language translation software applications to choose from that will meet the specific needs of a virtual project team.[16]

The following observations regarding the need for language translation in globally distributed virtual teams were made by Janet Astwood, an internationally-based IT Project Manager with experience in systems development in government, finance, and the publishing industries.

> "One of the advantages of hiring cross cultural/distributed teams is to acquire the best expertise in the field regardless of location, culture or language. On our IT projects our business language is English. Staff, not in the management role, has a mid to intermediate level in the English language, which we have found is more than sufficient to work effectively and collaborate in groups. Meetings are avoided as most staff members are not comfortable contributing their ideas verbally in English; instead communications between the teams is using real time messaging and shared documents. Requirements are prepared as images, videos and diagrams, with less reliance on written instructions wherever possible. This has

proven to be a successful model to involve all team members to participate and collaborate. Translation software enables staff to translate to/from their native language to communicate more complex ideas and concepts as needed.

I have 3 direct reports from China. I observed that they translated screens, email, documents from English/French to Mandarin. In stand-ups and meetings they would listen but rarely participate. They were competent and capable programmers and often their written English far exceeded speaking ability. If you find talent, available when you need them, then changing your communication strategy with more emphasis on real time messaging and images and less on in person verbal communications enables you and your team to bridge the language barrier.

We currently have several people on our team from Eastern Europe. Similar to China, we have observed that written English in Eastern Europe is good and they are reluctant to engage in meetings or verbal discussions. The staff, with mid-level English skills, communicates online in real time without translating. They do rely on the support of translation software for unknown words or phrases, however, in a work setting, the communication and the language used is generally brief, to the point and very repetitive. Technology continues to enable project teams to communicate faster with more clarity every day. Using screen capture tools and diagrams and translation software, our IT projects benefit from talent across cultures, languages and time zones."

File Sharing and Storage Technologies

Most virtual project teams now have access to and use some type of team space or groupware tools to move project files from one team member to another and to store the files. In some cases, however, the transfer and storage of files cannot be accomplished using these integrated tools. One such case is when some virtual project team members are located in areas where the internet infrastructure will

not support the use of highly integrated tools that require a high level of data bandwidth. Another case is when a virtual project is made up of team members from different companies and the separation of systems is common practice to ensure autonomy of company information. In these and other cases, the project teams need to rely on other means to distribute and store project information.

Email

Electronic file transfer via email is far and away the most common method used by both traditional and virtual project teams, even within teams that have access to and use team spaces and groupware applications. This is because of the simplicity and familiarity of using email. Tool use is always driven by ease of use, unless it is by management mandate.

Beyond the ease of file transfer, email also provides an opportunity to provide additional detail and context concerning the files that are moving from team member to team member. For example, for the sharing of a project requirements document, email provides the opportunity to describe that the team should be reviewing the non-functional requirement only and why (e.g., because the functional requirements are still being gathered). Email also provides opportunities to transfer files between two parties who have a specific need to know and to create a written record of the file transfer.

File Transfer Protocol

Even though file transfer protocols (FTPs) developed during the early days of the public internet, they are still used for transferring files between virtual project team members. FTP allows the user to transfer files between computers on a network or between a computer and an online archive. FTP also has the ability to navigate or search through a data archive and upload files of interest to the user.

Numerous blogs and articles by technology experts question the continued use of FTP when other modern technologies exist, such as HTTP.

Two reasons are commonly cited: The first is simply historical preference. The other reason has little to do with which technology is superior, but rather with how the technology is used. Unfortunately, technologists tend to become blinded by new technologies and are unable to see past features and functionalities. As the example in the box titled "Using FTP for IT First Responders" points out, which technology a team chooses should primarily be driven by usage needs.

Using FTP for IT First Responders

Justin Borges leads a virtual project team that consists of one co-located team in the northern part of his country, another co-located team that works in the southern part of the country, and six to 10 team members who are highly mobile and work anytime, anywhere. Borges's team works mergers and acquisitions for their company and are considered one of the "first responders" that go into a new company once it has been acquired. Their role is to get the new company up and running on the corporate IT systems quickly.

As Borges explains:

> We never know what technology environment we are going to have to work with. Some companies we acquire use the most modern technology, some companies use technology that is similar to our own, and many companies are very immature when it comes to technology. The majority of companies we work with are small, technologically immature, and located in geographically remote areas. Because we never know the technological landscape before going in, we exclusively rely on FTP to share files between our mobile workers who are on-site and the team members working at our major offices. FTP is simple, reliable, and requires relatively low bandwidth to transfer the files required to begin building a new corporate site for the acquired team.

Online Databases

The use of online databases has accelerated with changes to the internet driven primarily by cloud computing and Web 2.0. The increased use includes more adoption by virtual project teams. Online databases are designed to provide an organized mechanism for storing, managing, retrieving, and sharing project information from nearly any location and at any time of day. As illustrated in Figure 8.4, an online database can serve as the single source of information for the project team.

This single source of information is used to coordinate the daily workflow of the team regardless if team members are working from offices, from home, or from mobile locations. The team is able to make real-time updates to project information contained in an online database so everyone is working from the same up-to-date information.

The Wiki, now nearly two decades old, is one of the oldest online database solutions that is still in use today. Its original intent was to be "the simplest online database that could possibly work."[17]

Wikis don't have to be built by IT professionals with database expertise. Rather, the flexibility of the tool enables users to build a Wiki structure and with content that is most valuable to them. Many virtual project teams use Wikis as their project notebook or team intranet site where team members can add, modify, or delete pages or content within the pages using simple commands. If a project team needs to create its own knowledge management database and lacks the expertise or IT support to maintain a fully structured database, a Wiki site is a good option.[18]

Dedicated online databases are more prevalent in larger firms; smaller companies utilize online database services more often. This is due to a number of factors. Dedicated online databases require complex servers that must be operated and maintained by professional IT personnel. The cost of developing and operating a database for online applications can also be fairly expensive. Because online databases contain important business information, they are attractive to hackers. Therefore, the security risk is high, as is the cost of maintaining security.

Using online database services solves most of these problems for smaller companies that do not have the IT staff and budget to create, operate, and maintain their own secure database. However, the trade-off between cost and release of company and project information to the cloud has to be fully explored and debated. Security is always a risk, as is the possibility of data scraping by the hosting firm. Data scraping is a technique in which a computer program extracts data from human-readable output coming from another program. This is especially true for "free" online database services. (Remember the saying that if a service is free, the product is *you* and/or *your information*).

Social Networking Tools

For most people, social networking has become as much of a part of daily life as passing colleagues in the hallway or having a conversation over a cup of coffee. Social networking platforms have reached

Figure 8.4: Online Database as the Single Source of Project Information

such a high level of maturity in the public domain that they are being adopted in the business world as a means to enhance co-worker communication, collaboration, and relationship building. For virtual organizations and virtual projects, social networking tools have become a valuable means for keeping distributed people connected to one another.

Linda Church, project manager for a pharmaceutical company, uses social networking tools on every project she manages. "Our project teams are completely virtual with team members located around the globe," she says. "With our social networking platform, we instantly all know what's going on with the project, what's finished and what's in the queue, and how team members are doing on any particular day."

Social networking tools provide capabilities such as team member profiles, micro blogging, group discussions, private messaging, file sharing, and a searchable knowledge base created by team member interactions. The searchable knowledge base is accomplished by the use of tags. For example, if the team uses the tag "#budget," each time there is a conversation among the project team members about the project budget, anyone with access to the tool could search on that tag and the entire history of budget related conversations becomes available.

One of the most valuable aspects of social networking tools for virtual project teams is the ability to create communities where team members can share ideas as well as experiences relating to their personal lives. Communicating in this way builds relationships, trust, and morale among team members.[19]

The personal timeline is a very popular feature on public social networking tools. Some business social networking tools offer a similar feature that can be used to visually show a project timeline. The project timeline can be anchored by key project milestones, and team members can post work accomplished toward the milestones and artifacts relating to the work outcome, such as a screen shot of a graphical user interface design, for instance. This creates real transparency for project stakeholders and team members alike.

When it comes to problem solving, virtual project team members do not have the advantage of being able to walk the halls and get feedback on problems or challenges. This method is a form of physical crowd sourcing. Fortunately, social networking platforms can be used to simulate physical crowd sourcing virtually. Take, for example, a virtual project team member tasked with creating an outbound marketing plan, but she is new to the company. One of the challenges she will likely encounter is creating a list of clients to whom the marketing plan will be targeted. To tap into the collective knowledge and company intelligence, she can use virtual crowd sourcing techniques by posting a request for opinions on the highest potential clients using the social networking platform. In similar fashion, risk identification is a form of crowd sourcing activity that can be accomplished via social networking tools, as can the collection of opinions of risk severity.

No Shortage of Options

The wide proliferation of virtual work performed in our organizations today has been enabled by the availability of technology tools. The number of tools to choose from continues to grow, with no slowdown of tool development anticipated. One of the challenges for virtual project managers is staying current with viable technology options. A periodic inventory of technology options is always a good idea and provides an excellent opportunity for collaboration between a firm's IT department and project offices or organizations.

A technology inventory does not have to be, and probably should not be, extensive or exhaustive. It is best to focus on the most widely used tool options as the primary criteria for inventory inclusion. Other criteria, such as the most recently available technology or the technology with the highest number of integrated features, should be avoided. Table 8.1 provides an example technology option inventory of widely used tools by today's virtual project teams.

It should be noted that the technology inventory shown in the table is not exhaustive and may be outdated very rapidly. This is why we suggest performing a periodic analysis of technology options to use during the technology selection process. We describe that process in the next section.

Technology Selection

The principal parties who drive the use of technology on virtual projects within a company (senior managers, IT professionals, project managers) should select electronic technologies that best meet the needs and usage of the virtual project teams and that integrate with the current suite of tools used within the organization. Technology tools that complement the culture and that are appropriately selected based on user need should be chosen.

As we have noted a couple of time already, there is no ideal set of technologies for all teams. We reiterate it here due to its importance. Certainly, there are technologies (albeit basic technologies) that most all teams will benefit from. These technologies include the telephone, email, and scheduling systems. Since all organizations are unique, their virtual communication, collaboration, and relationship-building needs will also be unique. It is the role of the project manager—the team leader—to develop the strategy to match technology options to the communication, collaboration, and relationship-building needs of the virtual team. The project manager is also responsible for developing a technology needs assessment and resulting technology plan for the project.

Technology Assessment

By analyzing the various factors involved in a virtual project team's practices, the right technologies for a project can be selected.

The analysis does not have to be a complicated undertaking. A simple mapping of the various communication and collaboration factors that will be in play for a project to the various technology options

Table 8.1: Example Technology Inventory for Virtual Project Teams				
Category	Synchronous/ Asynchronous	Social Presence	Information Richness	Examples
Collaboration	Both	Medium to High	Moderate	Blackboard Collaborate Huddle Google Docs
Screen Sharing	Synchronous	Medium	Moderate to Rich	Join.me Free.Screen.Sharing Skype
File Sharing and Document Sharing	Asynchronous	Medium	Moderate	Drop Box Sharepoint Onehub
Conferencing	Synchronous	High	Rich	Telepresence Life Size Skype Facetime
Web Conferencing	Synchronous	Medium	Moderate to Rich	Go To Meeting WebEx Meeting Burner
Instant Messaging and Chat	Synchronous	Low to Medium	Lean	Jabber Google Skype Spark
Document Co-creation	Primarily Asynchronous	Medium	Moderate	Google Docs Microsoft Office Zoho Ether Pad
Social Networking (Internal)	Both	Medium to High	Moderate	Yammer Chatter Social Text
Meeting Management	Asynchronous	Low	Lean	Doodle Need to Meet Outlook Xoyondo
Bulletin Boards	Both	Low	Moderate	Marlin phpBB.com SMF Simple Machine Forum
Project Management	Asynchronous	Medium	Moderate	MS Project Primavera Basecamp Teamwork JAMA
Language Translation Services	Asynchronous	Medium	Moderate	Straker In What Language One Planet
Language Translation Software				Babylon LEC Prompt

Factors	Communication Technologies				Collaboration Technologies			Relationship Technologies
	Email	Blogs	Telephone	Video Conf.	White Boards	Team Space	Data Repository	Social Networking
Team Interaction								
Conversational	+		+	+				
Transactional	+	+				+	+	
Collaborative				+	+	+	+	+
Social Interaction				+				+
Time Differences								
Synchronous			+	+	+	+		
Asynchronous	+	+				+	+	+
Team Contexts								
Physical Infrastructure				−				−
Culture and Language			−					+
Team Size	−		−	+				+
Task Types								
Low Complexity								
High Complexity	−					+	+	

Figure 8.5: Virtual Project Technology Assessment Example

may be sufficient to formulate a plan. Figure 8.5 illustrates an example of a technology mapping outcome.

As the example demonstrates, the technology mapping shows the primary project team interaction factors that must be considered along with the various technologies options under consideration. Note that each organization will have its own unique set of virtual project factors and candidate technologies.

A simple +/- evaluation of whether a particular technology option either supports (+) or hampers (–) the various factors specific to the project will provide sufficient information to develop the project technology strategy.

By way of example, the information contained in Figure 8.5 represents the technology mapping results performed by Andreas Becker, a project manager in the automotive industry. Based on the mapping, the technology plan for Andreas's project included the following:

- The virtual project team is engaged in both conventional and collaborative transactions that are performed both asynchronously and synchronously. Therefore, email and telephone conversation, along with a shared workspace and a data repository such as an online database, are necessary to meet the team's needs.

- Physical infrastructure limitations in one or more geographies will likely prevent the use of more sophisticated technologies, such as video conferencing and complex groupware systems.

- A large team size and highly collaborative (high-complexity) workflow put constraints on the use of email communication and point to the need for audio conferencing capabilities that will support a large number of team members.

- The multicultural diversity of the large team requires social networking technologies to assist in building the relationships among team members who will likely never meet in person.

Two additional factors must be considered in the development of a technology strategy: technology maturity and technology overload. Organizations are often quick to adopt new technologies in hopes that they will help them to improve their communication and collaboration practices. However, new technology often brings both a user learning curve and technological bugs that may significantly hamper a virtual team's effectiveness instead of improve it. When this happens, teams eventually revert back to tried-and-true technologies, leaving the organization with a significantly low return-on-technology investment ratio.

In many firms, internal IT organizations bring a plethora of tools to help with team communication and collaboration. After all, this is the traditional mission of most IT organizations. The problem with this approach is that most times virtual project teams end up with more technology options than they need or that are useful. This problem points to a misalignment of goals between development organizations and the IT organization. To prevent technology overload, the goals of a firm's IT organization must align to the needs of the project teams so that technology selection is indeed driven by virtual team usage and need.

Increasing Technology Usage

Project team members who are new to the virtual setting are often reluctant to jump in with both feet and begin using technology to communicate and collaborate with teammates. Rather, many people who are new to virtual work (and veterans) often resist technology changes. A number of fundamental practices should be put in place to ensure the widespread usage of technology on virtual projects.

First, adequate training has to be supplied to all project team members when a new technology is deployed. Unfortunately, it has become common practice, especially in larger companies, to push new technologies to virtual teams without providing adequate training on how to use the technology effectively. We have all been in project settings where valuable time and resource effort is burned fussing with trying to get technology to work. We suggest that project managers take a hard stance on the need for adequate team training for all project members *before* new technology is deployed into the operational team environment.

The project team should develop a communication and collaboration charter as part of its team formation activities to specify how the team plans to communicate and collaborate. The charter should include team communication and collaboration rules, norms, and behaviors expected of each team member as well as what forums will be set up and how information will be shared and stored. The charter also should include the expectations for technology use in performing team communication and collaboration.

It seems like an obvious statement, but project managers must ensure that all team members have access to the technology used on a virtual project. People in remote sites may have constraints, such as internet bandwidth limitations, that may prevent the use of some technology options.

It is the responsibility of virtual project managers to ensure that the appropriate technology is chosen and that the team is adequately prepared to use that technology to collaboratively perform their work. This responsibility includes continually monitoring the use of technology to ensure that it remains an enabler and does not become a roadblock to progress.

Notes

1. Parviz F. Rad and Ginger Levin, *Achieving Project Management Success Using Virtual Teams* (Plantation, FL: J. Ross, 2003).
2. John Tuman and Paul McMackin, "Project Management for the Twenty-first Century: The Internet-Based Cybernetic Project Team." In *Procceedings of the 28th Annual Project Management Institute Seminars and Symposium Proceedings*, Chicago, IL 1997.
3. Deborah L. Duarte and Nancy Tennant Snyder, *Mastering Virtual Teams: Strategies, Tools, and Techniques that Succeed* (San Francisco, CA: Jossey-Bass, 2001).

4. "Global Project Development—Moving from Strategy to Execution." *Business Week* (2006).

5. Cristina B. Gibson and Susan G. Cohen, *Virtual Teams that Work* (San Francisco, CA: Jossey-Bass).

6. Trina Hoefling, *Working Virtually: Managing People for Successful Virtual Teams and Organizations* (Herndon, VA: Stylus, 2003).

7. www.wiki.answers.com.

8. Keith Ferrazzi, "To Make Virtual Teams Succeed, Pick the Right Players," *Harvard Business Review*, December 18, 2013. https://hbr.org/2013/12/to-make-virtual-teams-succeed-pick-the-right-players.

9. Susan B. Barnes, *Computer-Mediated Communication* (New York, NY: Allyn and Bacon, 2003).

10. Hoefling, *Working Virtually*.

11. Sharmila Pixy Ferris and Maureen C. Minielli, *Technology and Virtual Teams* (Hershey, PA: IGI Global, 2004).

12. Ferris and Minielli, *Technology and Virtual Teams*.

13. Hoefling, *Working Virtually*.

14. Robert Jones, Robert Oyung, and Lise Pace, *Working Virtually: Challenges of Virtual Teams* (Hershey, PA: CyberTech, 2005).

15. Ferrazzi, "To Make Virtual Teams Succeed, Pick the Right Players."

16. Celia Rico Pérez, "Translation and Project Management," *Translation Journal* 6, no. 4 (2002). http://translationjournal.net/journal/22project.htm.

17. http://wiki.org/wiki.cgi?WhatIsWiki.

18. Michael Klynstra, "Social Media Meeting Project Management," August 21, 2012. http://www.geneca.com/social-media-meets-project-management/.

19. Klynstra, "Social Media Meeting Project Management."

9

SUSTAINING VIRTUAL PROJECT SUCCESS

Much of the pressure to succeed in managing a virtual project falls on the project manager. However, consistent and sustainable success is enabled by critical organizational elements that are the responsibility of a firm's senior leaders. In a recent study conducted by the American Productivity and Quality Center, senior managers of virtual enterprises who were asked if they are responsible and accountable for creating and implementing strategy for their firms, all responded that they do in fact own that responsibility. When the same senior managers were asked if they are responsible and accountable for *execution* of the strategy, however, answers were generally mixed. Some fully believe they are responsible for execution success, while others believe the responsibility lies with middle managers and virtual project managers.[1]

The same study highlights the difference in philosophy among senior managers of firms considered leading virtual enterprises. Senior managers there believe they are responsible and accountable for both setting strategy for their companies *and* ensuring the necessary organizational changes and behaviors are driven across the enterprise to enable successful strategy execution by virtual project teams.

There is significant evidence that the frustrations and challenges project teams encounter in other virtual companies are due in large part to the failure of senior managers to lead in the removal of the execution barriers unique to virtual organizations. We define *virtual execution barriers* as fundamental organizational structures, power bases, and

behaviors that prevent companies from effectively operating in the virtual project environment. A series of organizational, operational, and philosophical changes must occur to prepare and enable an organization to move from a locally focused model to a geographically distributed model. Without these fundamental shifts, effective virtual project execution is severely challenged. The most common and significant barriers are the following.

- Organizational structures and performance measures that limit the collaborative team dynamics that are necessary on a virtual project.

- Differences in culture—country, company, and functional—that are not characterized, understood, and assimilated into the organization.

- A development model that does not support the highly collaborative and interdependent nature of distributed project team activities and promotes significant time delays.

- Skills and competencies of project managers that have not kept pace with the more comprehensive set of skills and competencies needed to be successful in the virtual environment.

In order for virtual project teams to begin operating more effectively within a highly distributed work environment, senior managers must step beyond setting strategy and become personally engaged in enabling virtual project execution success. Specifically, senior managers of geographically distributed companies need to establish the right structures and performance measures to foster a highly collaborative and distributed project environment. The managers must drive all changes necessary to deemphasize strong organizational silos that create collaboration barriers and must also change individual performance measures and rewards for their middle managers and team leaders to ones based primarily on achievement of team goals and secondarily on individual and departmental goals.

As a company expands its activities across the globe, it rapidly becomes a multicultural entity. Senior managers must work to converge national, functional, and organizational cultural aspects of the organization and its workforce. Doing so aligns the enterprise to a common vision, company value system, and way of doing business. The converged cultures manifest themselves as the project culture for each multinational project team.

Product and service development in a virtual environment requires a high degree of horizontal collaboration, synchronization, communication, and integration. Senior managers of virtual organizations must evaluate their current project execution model to determine if it is fully effective in a horizontally networked arrangement. If needed, they must drive changes toward a development model that is more collaborative and resistant to inherent time delays.

Senior managers must ensure that the role of their virtual project managers is appropriately defined to meet the broader based requirements to succeed in a virtual and potentially global environment. Along with appropriate role definition, senior managers must ensure that project managers possess the requisite knowledge, skills, and experience to lead a virtual team. The team leaders also must be empowered—have the responsibility and authority—along with senior management support—to effectively manage across the functions within an enterprise.

This chapter details the primary organizational changes that must be instituted by senior leaders of an enterprise to achieve consistent and sustainable virtual project success.

Changing Organizational and Team Structures

Many best-in-class virtual organizations have changed their fundamental thinking about how their organizations are structured. They have implemented approaches that minimize organizational hierarchies while de-emphasizing departmental silos and adopting more flexible organizations with horizontal networks of personnel and resources. This transition was spearheaded by many companies in the high-technology industry to take full advantage of their internationally distributed workforces.[2] This fundamental change in thinking is based on four foundational success factors for virtual organizations and their virtual project teams:[3]

1. Virtual project managers need direct access to the senior management leadership of the enterprise.

2. Successful execution of virtual projects is dependent on effective management of the interdependencies and interfaces among the distributed team members.

3. Geographically distributed virtual project members need timely and complete access to all important technical and business information that directly impacts their project.

4. Decision making power must shift from the top of the organization to the virtual project team within clearly defined decision boundaries.

Traditional, hierarchical organizational and team structures create a direct barrier to these four virtual project success factors. To ensure alignment between a firm's business strategy and its virtual project execution output, senior managers and virtual project managers must have direct access to one another. They must, in fact, work together as a leadership team. Traditional hierarchical organization structures normally prevent this from occurring due to the layers that exist between senior executives and project managers. The greater the number of layers between project managers and top-level management, the greater the probability of misalignment between strategic goals and execution output. This is a result of the lack of direct communication between executives and project managers as well as the skewing and diluting of communication as it passes through the layers of the organization.

Another complexity that can impact hierarchical organizations is *agency theory*. This situation occurs when there is a misalignment of goals between a manager (the "principal") and an employee (the "agent"). This is typically prevalent in organizations with strong functional silos. Agency occurs when functional managers design goals that provide the greatest benefit for their functional organization, with the strategic goals of the company being a secondary consideration.[3] If virtual project managers are reporting directly to functional managers and do not have direct access to senior managers, many times the project team drives to achieve the goals of the functional organization, not the goals of the overall business.

Since project managers within a hierarchical structure are most times contained within a functional silo, it is very difficult for them to reach across the organization and drive cross-discipline and cross-geographical collaboration in a virtual environment. A hierarchical structure can force project decisions to move beyond project managers to the functional managers of the organization. When a decision crosses organizational boundaries, the decision must move to the appropriate organizational function and down the chain of command within a silo. This method of decision making in a distributed organizational setting is very ineffective and inefficient because of the time it takes to reach a decision.

The same structural barriers exist when it comes to accessing technical and business information. Team members normally have access to the functional specific data and information contained within their organizational silo; however, seldom can they directly access data and information contained within another functional silo. The project manager, therefore, must resort to the same chain of command approval path just described in the paragraph above and will realize the same result—delayed or forbidden access to critical data and information.

Finally, hierarchical structures are power driven by design, with those at the top of the organization possessing the most positional power. Virtual project managers possess little positional power, which creates an execution barrier that makes it difficult for project managers to lead and influence a widely distributed set of project stakeholders. This situation normally leaves functional managers, who are more disconnected from the project and who are not familiar with the daily execution activities, in charge of project decisions.

To address these issues, best-practice virtual organizations have moved to organizational structures that are flatter, with fewer layers of management. An example of a flat organizational structure that works well for virtual organizations is illustrated in Figure 9.1.

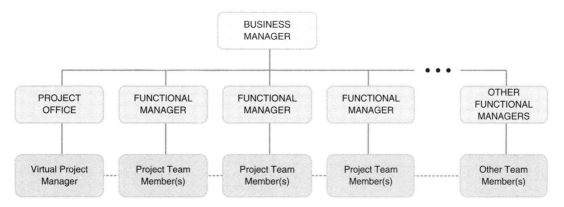

Figure 9.1: Example Virtual Organizational Structure

In this lattice or matrix structure, project managers have easy access to the top management of the organization. This access creates a direct communication channel and collaborative arrangement between senior leaders and virtual project managers. This communication linkage is a critical element in setting and maintaining alignment between business strategies to virtual project execution outcomes because it fosters direct communication, which leads to an aligned community.

A lattice structure also results in a de-emphasis of functional silos, with the various organizations now owning a shared responsibility for project success. Each of the functional managers must invest in the project by providing resources (people, money, equipment, or materials). The return on investment is dependent on the successful achievement of the project goals. Therefore, participation in cross-organization collaboration is in the functional managers' best interests.

Also of critical importance for success in a highly distributed virtual environment is that decision making power shifts to project managers. (See Chapter 6.) When a flat organization structure is instituted, the balance of power moves away from functional managers and to virtual project managers. This gives project managers greater decision making power and more degrees of freedom to operate.[4] Transfer of power enables virtual project managers to distribute decision making power to the local level, where people with the best information and most knowledge about a specific project situation perform their work.

Finally, a flat organization structure allows project managers to execute their primary role in leading the virtual team—integrating the work activities and work output of the distributed team. The flatter organizational structure enables project managers to work horizontally across functional disciplines of the enterprise. In doing so, they drive the creation of the holistic solution that directly contributes to the realization of the firm's strategy. (See Chapter 2.)

Modifying Virtual Project Team Structures

A change in the way a firm organizes its project team structure may be necessary to ensure maximum collaboration is occurring in the virtual environment. Like hierarchical organization structures, hierarchical team structures also stifle cross-team collaboration and communication. The project team structure therefore must be flattened to create a network of specialists and enable the integration of work flow and work output across the distributed team. Communication and collaboration should occur horizontally across the team, not vertically through a team hierarchy.

The project core team structure is the most common team structure found in companies that operate in a virtual project environment. (See Figure 9.2.)

Figure 9.2: Virtual Project Core Team Structure

Figure 9.3: Core Team Collaboration and Communication Triangulation

The virtual project core team is the cross-discipline leadership and decision making body of the project. Its members are responsible for ensuring that both project and business goals are achieved.[5]

The virtual project core team consists of the team leaders who represent the organizational functions and provide leadership for the delivery of their function's element of the product or service under development. [6] The core team must become very cohesive and be willing to share responsibility for the success of the project. As Figure 9.2 illustrates, the project specialists within each of the geographies are only one level removed from the virtual project manager.

The core team structure is highly integrated, meaning there is joint consideration of trade-offs, decisions, and problem resolution among members of the team. Coordination and communication within the core team occur both horizontally and vertically. Figure 9.3 demonstrates the triangulation of collaboration and communication that takes place on a core team. Directions, decisions, and cross-team issue brokerage comes from the virtual project manager, while cross-team communication and work coordination occurs between the on-site project team leaders. Status, decision consultation, and issue escalation flows from the project team leaders to the project manager.

Each member of the core team must be committed to the success of the other members on the team. A primary responsibility of virtual project managers is building a trusting, cohesive core team and leading them to mutual success by way of project success.

Modifying the Project Execution Model

Time is never on the side of virtual project managers. Time always presents a multitude of challenges that easily cause delays that are counted in days, weeks, and occasionally in months. Virtual projects face a loss of time efficiency due to their inherent complexity. Difficulties coordinating work and driving effective communication across multiple time zones is well known and documented. In addition to time zone challenges, however, time also presents a more perplexing and potentially dangerous challenge for virtual project managers. We are referring to time lapses between project activities, team deliverables, and project milestones caused by geographical separation of work. If not managed carefully, the lack of efficient time management by the project manager and the team can have an adverse impact to the project schedule.

Although even the newest virtual project managers learn quickly how to cope with their project's specific challenges, time lapse challenges are not as evident because much of the work and communication is happening behind the scenes.

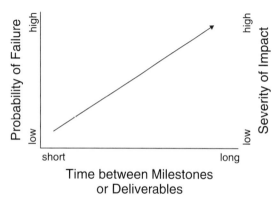

Figure 9.4: Increasing Risk with Increasing Time between Project Events

Many times time lapses in work outcomes do not become evident until a project moves outside its success control limits or until it fails outright. (See Chapter 4.)

Past analysis has demonstrated that the greater the time between project milestones, team deliverables, hand-offs, or cross-team touch points, the greater the risk that projects will experience significant schedule delays. (See Figure 9.4.)

Virtually distributed team members perform much of their work and create their project deliverables based on a set of requirements and a series of assumptions that are made during the project's planning stage. When requirements change or assumptions are found to be false, they are typically discovered during hand-offs of work elements or during integration activities.

With a decrease in direct communication, lack of informal meetings, and limited synchronous collaboration among distributed team members as compared to traditional project teams, it is rare that missed changes in requirements or incorrect assumptions are discovered through communication and collaboration. Rather, they are found when inconsistencies between what was delivered and what was expected are discovered.

The result, of course, is rework. Any time rework occurs, productivity decreases, time lapses occur, additional money has to be invested, and the project goals are compromised. To combat this negative impact to the project timeline, best-practice virtual organizations adopt an accelerated delivery mechanism that decreases the amount of time between project deliverables and hand-offs. We refer to this mechanism as a *rapid delivery model*. (See Figure 9.5.)

In a rapid delivery model, work output is decomposed into small deliverables, which can be worked on and completed within a short period of time. As shown in Figure 9.5, each team has a series of deliverables that are separated by two or three

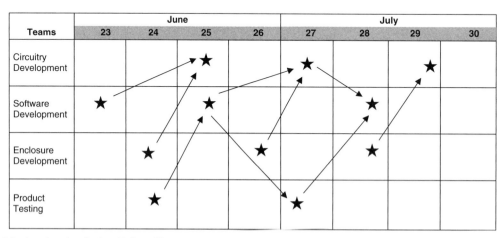

Figure 9.5: Accelerating Project Deliverables

weeks in duration, thus accelerating the integration process. With constant focus on rapid development and hand-off of deliverables among team members, it becomes difficult for the work of a virtual project to veer too far off course before the problems and miscommunications are discovered and corrected.

This process provides the team the ability to test assumptions in an accelerated manner and potentially reduce the amount of rework. This step-by-incremental-step model is highly effective and rather standard on virtual projects. The model works because it forces team members to communicate and collaborate on a continuous basis throughout a project life cycle. When this model is utilized, the need to create a project map in addition to a project schedule increases (see Chapter 2), as this tool becomes the team's focal point for planning and managing rapid development of its deliverables.

Changing Behavior by Changing Rewards

How well a team works together is, of course, critical to project success, especially so for virtual projects. In Chapters 5 and 6, we emphasized the importance of establishing a "One Team" philosophy while building a virtual team during the initiating and planning stages of a project. To sustain the One Team philosophy throughout the remaining stages of the project, a firm must modify the recognition and reward system for its virtual teams to emphasize and reinforce team performance over individual performance.

All teams, whether traditional or virtual, share at least one common attribute: To be highly effective, they require a supportive reward system where personal success is dependent on team success.[7] Effective execution in a virtual project environment requires a high degree of collaboration among team members; therefore, the reward system for virtual teams must put a premium on collaborative success. Team performance measures must be designed to include working across geographical boundaries, successful integration across cultures, sharing of

information, completing interdependent deliverables, sharing critical information to support virtual teamwork, and meeting team objectives. Because of these important team-based factors, this requires rewards that are based on *team results* instead of *individual effort*.[8]

The challenge is to create a fit among the characteristics of the individuals on the project team, the goals of the virtual project, and the characteristics of the reward system. Because teams can differ greatly in mission, structure, processes, and member makeup and distribution, no one reward system design will be universally effective. Each organization must design its own system. Doing so, unfortunately, is not a simple task. A common approach is to design a reward system that combines both skill-based and performance-based rewards, both of which are heavily influenced by the achievement of the project goals.

Skill-Based Rewards

Skill-based rewards are used on virtual projects to motivate the development of functional expertise of virtual project managers as well as team members. In addition, skill-based rewards are used to focus on the development of cross-functional knowledge and virtual team operating skills that are needed to participate as effective and capable team members of the virtual project and drive virtual communication and collaboration.

Depth of expertise in one or more functional areas utilized on the virtual project is often critical to execution success. Skill-based incentives reward individuals for developing deeper levels of expertise within a knowledge area and help to retain the technical expertise within the organization. Typical implementations of skill-based rewards involve defining multiple levels of expertise within a knowledge area (such as project management, engineering, marketing, or manufacturing), and when individuals demonstrate that their competency exceeds their current level, they are promoted to the next higher level.

Performance-Based Rewards

To ensure that skill-based rewards are impacting team results, virtual project team individuals must prove to the project manager and senior management that they have demonstrated that the skills within their functional expertise level were used to help further team collaboration and contributed to the success of the project on which they participated. Team-based performance achievement must be a defining criterion for promotion to the next level. This is the basis of performance-based rewards.

We must discuss two important points concerning performance-based rewards. First, bonus plans do a better job of motivating team members than skill-based pay raises. Second, team members assign higher credibility to objective performance measures based on team achievements.[9] Therefore, organizations that tie rewards to quantifiable measures can more effectively shift team member behavior from an individual focus to a team focus than organizations that take a more subjective approach, such as utilizing a manager's rating system.

The challenge in using performance-based reward systems for virtual project teams is that the reward system must be specifically designed to support a particular team. The team reward system must use measures and metrics that demonstrate successful team performance and tie the rewards to the achievement of project goals.

An additional challenge is the timing of the rewards. The most common practice for rewarding team performance is to recognize and reward team members at the conclusion of the project. Often, however, team membership changes over the duration of the project. Therefore, it becomes difficult to gauge the contribution of members who participate in but are not part of the effort at the conclusion. One approach to address this challenge is to implement a phased approach to giving rewards. As team members achieve goals through various project phases, incremental rewards are given to those who participated and are deserving of reward in each phase.

Management Rewards

Recognition and rewards for senior and middle managers offer challenges as well. Specifically, we are referring to the organizational managers who provide resources and other support and services to virtual project managers and teams. The rewards and recognition model shifts significantly as an organization transitions toward a network structure necessary to enable effective virtual collaboration. As we discussed in earlier chapters, one key characteristic of the collaborative approach is one of purposely driving project related decision making from organizational managers to virtual project managers.

Senior leaders must evaluate their middle managers based on the shift to the new model. Functional and department managers should be rewarded not only for their efforts in building strong functional teams, but also for their roles in supporting virtual projects. Specifically, rewards should be based on how well they provide mentoring and coaching and whether they help remove obstacles, provide creative input, and guide team member skills development—all of which lead to organizational value and long-term virtual project success. Importantly, virtual project managers should be given the opportunity to provide performance feedback for a firm's middle managers.

Senior leaders should be evaluated and rewarded on how well the organization has transitioned to a virtually distributed model. Senior leaders must design an organization that effectively operates in the virtual project environment. Then they must pay careful attention to establishing the opportunities for virtual project teams to perform successfully.[10] Doing this also includes ensuring that the rewards and recognition for organizational managers are properly aligned and administered in order to contribute to the success of virtual project teams.

Promoting Cultural Awareness

Organizations that utilize virtual teams on an international basis will face workforce cultural diversity resulting from the acquisition of talent worldwide. The effectiveness of any internationally distributed virtual project team depends largely on the engaged participation of its team members from all geographies involved, where participation involves contribution of information, sharing of ideas, and involvement in the team decision process. Proper management of cultural diversity and intercultural interaction among team members is therefore critical to team success. Senior management of organizations must revamp the cultural vision of the enterprises to embrace and fully benefit from the increasing involvement of multicultural teams.

Laura Smith, executive director of U.S. Human Resources for Edelman Public Relations, a firm of 65 global offices and over 5000 employees, is an expert in managing an international workforce. She is an advocate for corporate cultural awareness programs and believes one critical element is needed for program success: "Lead and support it from the top of the organization."[11] Smith also suggests that senior managers consider adopting these factors as being important in cultural awareness programs:

- Ensure a clear strategy is in place when acquiring talent.

- Create networking opportunities to build an inclusive culture.

- Conduct ongoing cultural awareness training to promote understanding.

- Measure and evolve supporting business processes.

Employees who are more aware of and comfortable working in culturally diverse organizations are more prepared to recognize and act on global opportunities and are able to operate more effectively in a variety of cultural and business environments, whether traveling abroad or participating as team leader or a member of internationally distributed teams.

Various team leadership studies indicate that looking at the impact of emotional intelligence traits, especially the social skill trait, may be a key element to effective cross-cultural leadership.[12] Figure 9.6 illustrates this point.

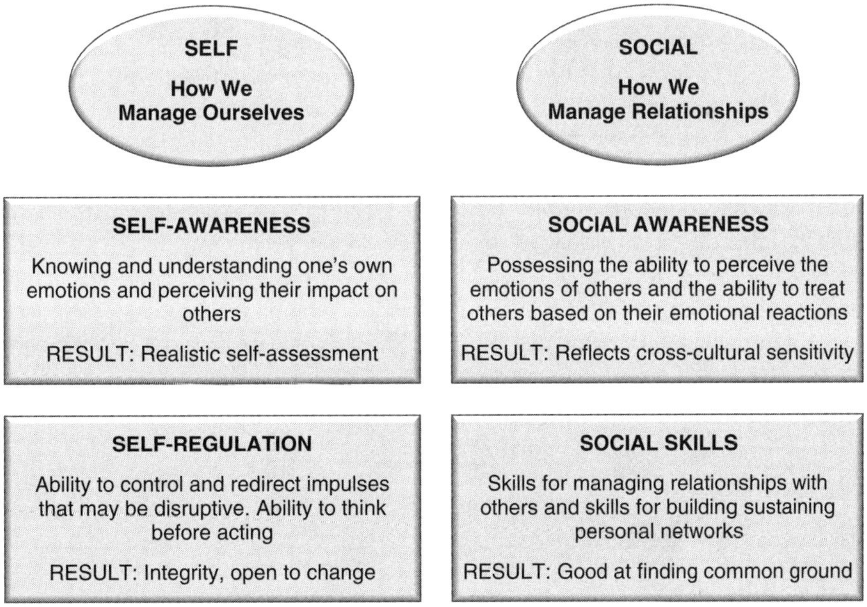

SELF
How We
Manage Ourselves

SOCIAL
How We
Manage Relationships

SELF-AWARENESS

Knowing and understanding one's own emotions and perceiving their impact on others

RESULT: Realistic self-assessment

SOCIAL AWARENESS

Possessing the ability to perceive the emotions of others and the ability to treat others based on their emotional reactions

RESULT: Reflects cross-cultural sensitivity

SELF-REGULATION

Ability to control and redirect impulses that may be disruptive. Ability to think before acting

RESULT: Integrity, open to change

SOCIAL SKILLS

Skills for managing relationships with others and skills for building sustaining personal networks

RESULT: Good at finding common ground

Figure 9.6: Emotional Intelligence and Cultural Awareness

Creating long-term relationships among multicultural team members, clients, and other members of globally distributed organizations is an excellent example for utilizing social skill. This result can be further characterized as proficiency in building relationships and networks and the ability to build excellent common ground and needed rapport with all personnel with whom virtual project managers must interact to perform their roles successfully.

Many scholars and practitioners agree that emotional intelligence capabilities need to be included in any cross-cultural training and development to assist virtual project team members and other employees in recognizing elements and nuances associated with varying cultural backgrounds. Additionally, virtual project managers should be selected based on their training and knowledge in specific emotional intelligence competencies, such as empathy, emotional poise, and self-control.

Developing Virtual Project Managers

It is rare when a person comes to the role of a virtual project manager fully qualified, possessing both the full set of skills and competencies required and experience. Successful virtual project managers constantly seek to learn and broaden their knowledge and experience to take on more complex and challenging work. It is therefore critical that senior leaders of virtual organizations create a learning environment that encourages virtual project managers to continually seek improvement and growth.

Noel Tichy, author and noted scholar, has stated that while management skills in the areas of finance, manufacturing, and marketing are important for organizational success, they are insufficient for effectively leading, planning, and sustaining organizational change and transformational initiatives.[13] We must keep in mind that all project managers are *change agents*. Through the performance and implementation of their projects, they transition the status quo within an organization from point A to point B; in other words, from current state to future state. Doing this requires a broad-based set of skills that goes beyond traditional management.

Project managers who manage virtual projects must be highly competent and capable individuals who possess the right skills to lead virtual project teams. However, success in the role goes beyond skills alone; they also must be highly *competent* individuals. *Competence* is defined as the knowledge, skills, and qualities that managers use to effectively perform the functions associated with management in the work situation.[14] A simple algorithm sums it up best:

$$\text{Competence} = \text{Knowledge} + \text{Skills} + \text{Personal Qualities} + \text{Experience}$$

Figure 9.7 was designed to characterize the necessary competencies and skills needed to succeed in managing virtual projects and leading distributed project teams. The information presented in the model is based on research by the authors and was derived from firms that execute effectively in highly distributed project environments. As the model demonstrates, much of the success of virtual project managers is driven by their behavioral and

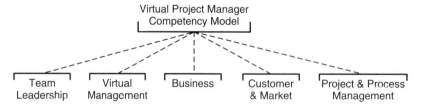

Figure 9.7: Virtual Project Manager Competency Model

human-oriented capabilities.[15] In the sections that follow, we provide an overview of each competency area.

Team Leadership Competencies

The foundational elements of effective team leadership apply whether one is leading a traditional team that is co-located at a single site or an international team that is highly distributed across continents. Virtual project managers must learn how to apply their team leadership capabilities in a virtual, and potentially multicultural, team environment. Imagine, for example, the challenges you would face trying to influence key managers critical to the success of your project who are located in other geographical locations with no face-to-face personal contact. This dilemma, no doubt, makes leadership more difficult for the virtual project manager.

Figure 9.8 summarizes the crucial team leadership skills for the virtual project manager.

Specifically, we are referring to the ability to lead cross-organizational, cross-geographical, and cross-cultural teams. Virtual project managers need to be able to build, coalesce, champion, and lead the team to create solutions that will satisfy the company's customers. The team leadership competencies fall into two categories: core leadership skills and augmented leadership skills. Team leaders who possess the augmented skills, in particular, are significantly more likely to be successful in virtual project environments.

Core Leadership Skills

The foundational elements of effective team leadership apply for domestic teams that are co-located in a single site or global teams that are distributed across multiple sites and countries. Success begins by applying the core principles of team leadership and then understanding how to extend these leadership principles for application in distributed team environments. The core principles of team leadership are listed next.

- Creating a common purpose
- Establishing team chemistry
- Building and sustaining trust
- Demonstrating personal integrity
- Empowering the team
- Driving participation, collaboration, and integration
- Communicating effectively
- Managing team conflict
- Making tough decisions
- Providing recognition and rewards

We described the core team leadership principles in detail in Chapter 3 and Chapter 5.

Augmented Leadership Skills

In addition to the core leadership competencies, several other important leadership skills need special attention due to their importance to virtual

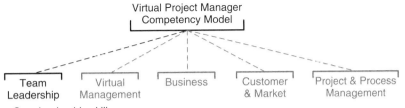

Figure 9.8: Virtual Project Manager Team Leadership Skills

team environments. These include influencing skills, prioritization skills, symphonic and systems skills, and political savviness skills.

Influencing Skills

In virtual team environments, team members rarely report directly to project managers. For this reason, project managers must become proficient in influencing team member actions. Project managers also need influencing skills to positively affect the actions and decisions of the senior management team, key partners, suppliers, customers, and support organizations. As John Maxwell states in *The 21 Irrefutable Laws of Leadership*, "Leadership is influence— nothing more, nothing less. If you don't have influence, you will *never* be able to lead others."[16]

Successful influencing involves gaining support for your team when needed, inspiring others to do their best, persuading others to follow your direction and coalesce around a common team purpose, and creating strong relationships. It is about moving things forward without pushing, forcing, coercing, or threatening. Influencing traits of a strong virtual project manager include being socially adept at interacting with others in any given situation, having the ability to assess all aspects of information and behavior without passing judgment or injecting bias, and being able to effectively communicate your point of view to change opinion or change course of action.

Prioritization Skills

The ability of virtual project managers to set and balance team priorities is one key indicator of success. The first step in setting priorities is checking to make sure the assumptions driving the project priorities are correct. It is one thing for team leaders to set the priorities; those priorities must be validated with the primary sponsors of the project. If the assumptions behind the priorities are incorrect, the priorities themselves may be incorrect. Once the priorities are validated, project managers should manage to the priorities. For example, if cost containment is the highest priority of the project, a project manager must be emphatic about staying within the financial constraints. If technological leadership is the highest priority, the project manager will need to keep the team focused on the technical aspects of the project and provide the necessary resources to ensure technological success.

Symphonic and Systems Skills

Symphonic skills involve achieving balance and optimization across multiple diverse, but related elements. It represents arranging the pieces together harmoniously with the objective of achieving a synergistic improvement. Being able to see the big picture, crossing boundaries, combining disparate elements, seeing broad patterns are all characteristics of symphonic skills. This ability normally resides in individuals with broad backgrounds, multidisciplined mindset, and a broad spectrum of experiences.

Systems skills involve the ability to assemble pieces together harmoniously, resulting in a synergistic improvement. People who have systems skills can see the big picture, combine elements into a new holistic entity, and see relationships between unrelated fields and broad patterns. Usually this ability resides in people with very wide backgrounds, multidisciplinary minds, and broad experiences.

Systems skills also include the ability to see relationships between relationships, which is also known as systems, gestalt, and holistic thinking.[17] Author Peter Senge presented systems thinking as a framework from which one can organize and understand events, behaviors, and phenomena that affect one another in the short- and the long-term. Those who apply systems thinking rather than linear thinking can see the dynamics that are reinforcing an event or limiting growth.[18]

Systems skills are also critical to virtual project managers. A system is a collection of parts that can be combined into an integrated whole to achieve an objective or entity. To understand this definition, you must understand what a functioning integrated whole is. For example, a bicycle is a functioning system. However, if you remove the handlebars, it no longer is a functioning bicycle. Those skilled in systems thinking can view projects and activities from a broad perspective that includes seeing overall characteristics and patterns rather than just individual elements. By focusing on the entirety of the project or, in essence, the *system* aspects of the project (inputs, outputs, and interrelationships), virtual project team leaders improve the probability of achieving the practical end solution and customer expectations.

Political Savviness

Organizational politics originate when individuals drive their personal agendas and priorities at the expense of a cohesive corporate agenda. Company politics are a natural part of any organization, and virtual project managers should understand that politics are a behavioral aspect of leading that they must contend with in order to succeed. The basis of organizational politics is really twofold: the desire to advance within the firm and the quest for power (usually in the form of controlling decisions and resources).[19]

Virtual project managers must actively manage the politics surrounding their projects to protect against negative effects of political maneuvering on the part of stakeholders and to exploit politically advantageous situations. In order to do this, project managers must possess both a keen understanding of the organization and the political savviness necessary to build strong relationships to leverage and influence the company's power base.

The most effective way to manage in the organization's political environment is to leverage the project stakeholders and powerful members of the network who can help achieve the project objectives.[20] The key is to avoid being naïve and to understand that not every stakeholder sees great value in the project. In our experience, virtual project managers who practice effective stakeholder management are more likely to succeed. Stakeholders come to the table with a variety of expectations, demands, personal goals, agendas, and priorities that many times are in conflict with one another. Project managers must rationalize and resolve these competing requirements by striking an appropriate balance between stakeholders' expectations and project realities. Team leaders, therefore, must be politically astute by being sensitive to the interests of the most powerful stakeholders and must, at the same time, demonstrate good judgment by acting with integrity.[21]

Virtual Management Competencies

Besides the normal challenges facing virtual project team leaders (time, distance, language, culture, and limited face-to-face interaction), another challenge is that, as in most projects, team members do not report directly to project managers. Leading teams of people who not only do not report to them directly, but also are distributed geographically create greater team leadership challenges. To deal effectively with this core principle of virtual projects, project managers have to develop and refine what we refer to as *virtual management skills*. Figure 9.9 shows a number of the more critical virtual management skills. Virtual project managers with these skills are at a significant advantage.

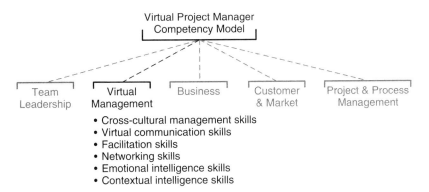

Figure 9.9: Virtual Management Skills

Cross-Cultural Management Skills

Competence at cross-cultural management is critical in leading teams in the highly distributed, multisite environment. Cross-cultural management skills are the ability to understand the behavior of people from diverse nations and cultures. These skills include awareness of cultures managers are directly involved in and understanding attitudes, differences, and behaviors. Such skills emphasize improving the interaction and working relationship among team members, management, and suppliers from all the cultures represented in the direct and broader team. To be skilled at cross-cultural management, we must examine our own biases and prejudices and, when possible, observe and learn from culturally proficient role models.

In regard to leadership skills and style, one approach may work in one culture, but not work in another. According to Nancy Adler, author, professor and consultant,

> Some researchers suggest that American approaches to leadership apply equally well abroad. However, most believe that leaders must adapt their style and approach to the cultures of the involved employees and clients. That is, they believe that leadership is culturally contingent.[22]

Regarding observed actions and behaviors in various cultures culturally skilled virtual project managers also are adept at:

- Listening for hidden communications in voice intonation and looking for non-verbal cues relative to facial expression, behavior and physical movement.

- Watching for blinders to cultural sensitivity in themselves and team members, such as stereotyping and projected similarity.

- Studying their team members and interpreting their specific cultural biases in regard to power distance, uncertainty avoidance, context, and perceptions of career success and quality of life.

Only through understanding and appreciating the unique characteristics of team members'

cultures can team leaders show members the respect and understanding that each person desires and deserves. Few team leaders ever obtain a complete level of understanding. However, this should not prevent us as team leaders from making our best efforts to sharpen our cultural skills, awareness, and behavior.

Virtual Communication Skills

Communicating virtually requires us to broaden our perspective and appreciation for the entire communication process due to the comprehensive challenges facing the exchange of meaning in a highly distributed, multisite environment. *Communication* is any behavior another person perceives and interprets as the understanding of what was meant. Communication includes sending both verbal messages (words) and non-verbal messages (tone of voice, facial expressions, behaviors). It includes consciously sent messages as well as subtle messages that senders are totally unaware of having sent. Communication, therefore, involves a complex, multilayered, and dynamic process through which people exchange meaning.[23]

Communication on a geographically distributed team is complicated by the physical separation of team members and the resulting reliance on technologies to facilitate team communication. Virtual team leaders need to develop skills in selecting the appropriate communication technologies given the tasks required, technical competence of the team members, and infrastructure capabilities within the geographies in which team members reside. Virtual team leaders must then become proficient in the use of communication technologies selected so they can teach other team members how to use them. Team leaders also must be able to use and model the appropriate communication etiquette associated with a technology.

Facilitation Skills

Simply put, *facilitation* is the act of assisting team members to reach their collective goals by helping to make team communication and collaboration easier and more effective. Good facilitation skills help to ensure that relationships among team members continue to develop and that ongoing communication and collaboration among team members is occurring as needed. Facilitation skills needed in a virtual environment go well beyond those needed for co-located project teams.

Communication and collaboration in a virtual team environment does not occur spontaneously at the onset of a virtual project, even if team members are familiar with one another and have worked together previously. The geographic and time separation among team members creates communication and collaboration challenges. Virtual project managers must therefore utilize facilitation skills to overcome the time, distance, and cultural barriers to stimulate and sustain effective virtual communication and collaboration within their teams.

Many aspects of successful leadership of a virtual project team covered in this text rely on strong facilitation skills on the part of team leaders to:

- Craft the project vision and common purpose,

- Establish the team norms,

- Create the team charter,

- Solve project-related problems,

- Build strong personal relationships,

- Reach good decisions,

- Manage conflict between team members,

- Identify and manage cross-team deliverables,

- Brainstorm new ideas, and

- Enforce team rules.

Each aspect of virtual team leadership just identified is critical to project success, and each requires facilitated discussion and collaboration among the virtual team members. Primarily, virtual project managers must provide the facilitation leadership.

Core facilitation skills include the ability to draw out varying opinions and viewpoints among team members, to create a discussion and collaboration framework consisting of a clear end-state and discussion and collaboration boundaries, to summarize and synthesize details into useful information and strategy, and to lead the adoption of technological communication and collaboration tools. Other facilitation skills that are beneficial include using personal energy to maintain forward momentum, being able to rationalize cause and effect, helping team members to establish one-on-one relationships, and keeping team members focused on the primary topics of discussion and collaboration.

Arguably, the most critical facilitation skill is project managers' ability to lead virtual meetings. Since a significant amount of communication and collaboration among members occurs in team meetings, virtual team leaders have to develop skills in planning and conducting remote meetings. Virtual team meetings will run the gamut from some face-to-face meetings, phone conferences, video conferences, internet-based data sharing meetings, or some combination of all of these. These meetings should involve pre-planning an agenda with time-boxed topics, sending any necessary materials to all members prior to the meeting, setting the meeting ground rules, facilitating the discussion to ensure a mutual understanding of all conversations, and periodically checking to see if quiet members understand the discussion and are fully engaged.

Utilization of the various technical tools available should enhance good facilitation practices by enabling team members to share concepts, merge information, and formulate new ideas. Technological tools, however, will not make up for poor and improperly facilitated meetings.[24] Project leaders are responsible for facilitating effectiveness and relying on technology tools for efficiency.

Networking Skills

The ability to network successfully across worldwide hierarchical and organizational boundaries is a tremendously useful skill, given that customers, senior managers, team members, and other critical parties are dispersed across multiple distances, sites, and countries in a virtual organization. Virtual project team leaders first must know how to determine the organizational landscape and who is in it. Doing this involves effective stakeholder identification and political mapping capabilities. Team leaders then must be able to use and extend this knowledge to develop the ability to choose the right mode of communication to address customers, senior management, team members, suppliers, and others.

As a keystone member of the project network (see Chapter 6), the virtual project manager must possess the ability to connect other individuals within the network needing to communicate and collaborate with one another. The strength of the project network is dependent on direct interconnections among team members who are virtually separated and lack the ability or knowledge to create the connections on their own. Networking skills give virtual project managers the ability to do this.

From a team perspective, effective networking skills gives virtual project managers the ability to create a sense of urgency in team members who may be isolated from the rest of the team or are being pulled toward other competing priorities. Networking skills are required to assist these team members in feeling like they are part of the team and to align their personal work priorities to those of the project.

Emotional Intelligence Skills

There exists a strong argument that the Intelligence Quotient (IQ), which traditionally has been the measure of intelligence, ignores key behavioral and personality elements. Beyond IQ, success depends on the awareness, control, and management of our own emotions, while influencing the emotions of others around us. This forms the basis of the concept of Emotional Intelligence (EI). As leaders of virtual enterprises have searched for the most critical leadership competencies, they have learned that EI contributes to as much as 90% of the differences between star performers and average performers.[26] Additionally, research by the Center of Creative Leadership found that the primary causes of derailment in executives' career aspirations

involve deficits in emotional competence, such as difficulty in handling change, not being able to work well in a team, and poor interpersonal relations.[27]

Daniel Goleman, author, psychologist and science journalist, describes emotional intelligence as "managing with heart."[28] His groundbreaking book, *Emotional Intelligence*, redefined what it means to be competent as a leader and describes EI skills as critically important in virtual team environments where face-to-face interaction and involvement is limited. Being acutely in tune with and sensitive to emotions and emotional responses of team members is critical to building a high-performing team.

EI skills consist of personal competence and social competence. Personal competence involves both self-awareness and self-management, where self-awareness is the ability to accurately perceive one's own emotions and moods in the moment and understand one's tendencies in various situations. Self-management is the ability to use awareness of emotions to stay flexible and direct one's behavior positively. Thus, self-aware project managers stay on top of their own reactions to team members and others and manage their own emotional self-regulation to think before acting or reacting.

Social competence includes social awareness and relationship management skills that enable project managers to understand others and manage relationships. Social awareness is the ability to accurately pick up on the emotions of others and to understand what is at the root of the emotions. In essence, socially aware project managers are empathetic and can understand and react appropriately to the emotional needs of others. Relationship management is the ability to use personal competence and social competence to recognize both one's own emotions and those of others to manage interactions successfully. This capability forms the foundation for the bonding and building of long-term personal relationships over time.

Contextual Intelligence Skills

The contextual environment in which project managers operate in virtual settings is increasingly complex. The environment is continually evolving and is both dynamic and turbulent. Decisions must be made quickly and must be useful and practical. Project managers able to perform successfully given these challenges have a high degree of contextual intelligence.

Context is the setting in which events occur. It consists of internal and external factors surrounding the circumstances of the event. Understanding context can directly impact how one responds to the events and to transform data into useful information. Matthew R. Kutz, author of *Contextual Intelligence: An Emerging Competency For Global Leaders*, defines contextual intelligence as "the ability to quickly and intuitively recognize and diagnose the dynamic contextual variables inherent in an event or circumstance, which results in intentional adjustment of behavior in order to exert appropriate influence in that context."[29] It results in integrating and diagnosing information while exercising and applying knowledge pertinent to the contextual situation.

Contextual intelligence skills are "innate [abilities] to synthesize information quickly and effectively," according to researchers Erik Dane and Michael Pratt. People with these skills are astute at detecting attitudes, motivations, and resistance of parties involved in specific events.[30] Team leaders who possess a high degree of contextual intelligence are able to cognitively and intuitively assimilate the various data and observations surrounding an event and convert this new understanding into information useful to making decisions.

Business Competencies

Strategic thinking and business fundamentals are key competencies needed for project managers leading and managing complex and critically important projects. This is true whether the project is a traditional or a virtual one. In many enterprises today, strong business competencies are required to fulfill the project manager role, including the ability to develop a compelling project business case that supports the company's business goals and strategies, the ability to execute the project from the business perspective, and the ability to understand and analyze the financial aspects of a project. (See Figure 9.10.)

Project managers may be called on periodically to apply these skills and capabilities during a project. These business skills are definitely enhanced through virtual project managers' prior experience and exposure to international markets and customers.

Business Fundamentals Skills

To be successful from a business perspective, virtual project managers must possess sufficient business skills to understand the organization's business model and financial goals. They must be able to utilize economic, financial, and organizational data to build and document the business case for their projects and be proficient in business terminology when communicating with senior managers and other business stakeholders.

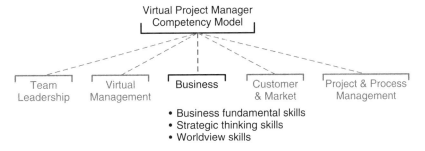

Figure 9.10: Business Skills

Virtual project managers must have a working knowledge and some proficiency in business fundamentals including ability in financial analysis and accounting, international management, political issues, law and ethics, resource management, negotiation and communication, and management of intellectual property. In addition, project managers may need to possess a working knowledge of the local and international economics in which the project is operating.

Strategic Thinking Skills

Virtual project managers must think strategically to align projects to the organization's strategic objectives. A part of strategic thinking involves a basic understanding of the industry in which the business operates. Knowledge of industry trends, competitors, and supply chain implications is a fundamental part of keeping projects viable.

Worldview Skills

Being proficient in global business also means possessing a worldview. Possessing *worldview* skills involves having an awareness of the global environment, including social, political, and economic trends. Managers of multinational projects must be able to apply their worldview knowledge, skills, and competencies to consistently succeed while participating as a critical member of a firm's global business.

Globalization is driven by a set of forces that have operated interdependently throughout recent history. (See Chapter 1.) Knowledge of the three primary globalization forces—economic forces, political forces, and technology forces—provides virtual project managers a greater contextual understanding of the environment in which they operate. Project managers' worldview capability is, of course, significantly enhanced through personal experience and exposure to international markets, cultures, and team members.

Customer/Market Competencies

For virtual project managers, customer and market competency involves having a thorough understanding of the industry and market in which their companies are doing business and how the product, service, or other capability they are creating is used. The better virtual project managers and their team can align the capability with customers' needs, the greater the potential for customer satisfaction and project success. Figure 9.11 highlights the skills involved in customer and market competency.

Product or Service Knowledge

At a minimum, virtual project managers must understand customer needs and desires that are pertinent to the new product, service, or other capability under development. This understanding requires that project managers have sufficient technical knowledge to recognize how the needs

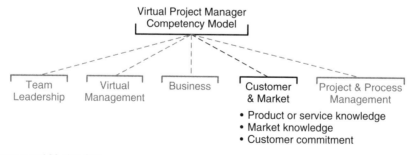

Figure 9.11: Customer and Market Skills

can be met by the capability and how to integrate the elements of the design and development into a successful solution for customers.

Market Knowledge

As a company grows, scales, and expands, its customer base and the markets it sells in or serves become more diverse. These changes may require virtual project managers to embark on a continual learning path to stay abreast of various markets. We do not mean that project managers need to be the customer, market, and industry expert on the team. This is not the role of the project manager and would most likely consume most, if not all, of a manager's time and energy. Rather, virtual project managers must know how to tap in to this type of expertise within the enterprise to maintain a general level of knowledge and to bring pertinent information to the virtual project team.

Customer Commitment

Virtual project managers must be consummate customer advocates for the projects they manage. This means being skilled in voice-of-the-customer. The "voice of the customer" is a process used to capture the requirements from customers, both external and internal, in order to provide customers with the best product and service quality. These techniques ensure customer needs and desires are reflected in the final capability. As customer advocates, project

managers must understand how their customers define quality and how it should be measured.

Process and Project Management Competencies

Process and project management is the final element of the virtual project manager competency model. As shown in Figure 9.12, project managers of both traditional and virtual projects must be trained and competent in the core processes of the firm they are serving. In addition, they must possess the fundamental project management knowledge, practices, methods, and tools to manage their projects to a successful outcome. Possessing these competencies raises the probability that project managers will gain and hold the confidence and trust of the project team members, project stakeholders, and customers.

Process Proficiency

An important aspect of this core skill set is that of possessing a solid working knowledge of the specific processes and practices of the company. Knowing how things get done, the policies and procedures that must be adhered to, and who must be involved and approve various aspects of the project is critical for the successful completion of every project. If it is a product-based company, for example, project managers must be thoroughly familiar with the firm's new product design, development, and market launch processes to ensure that

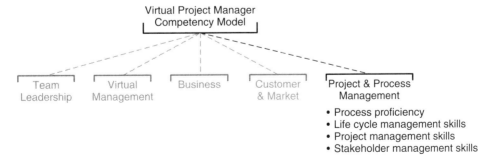

Figure 9.12: Process and Project Management Skills

team effort adheres to management's requirements and expectations as to how products are designed and built.

Life Cycle Management Skills

A challenge in leading a virtual project team is ensuring that all team members are following the same processes and methods and are using the same tools, when appropriate. A foundational element in driving process consistency across the virtual team is to ensure that all members adhere to a common life cycle for the coordination of work activities from ideation to project closure. The ability to effectively use a life cycle to manage a virtual project team's work will help to drive common language and terminology, establish a common cadence of activities, and provide common decision and synchronization points throughout the project cycle.

Project Management Skills

In addition to being skilled in the core project manager competencies required (e.g., Project Management Body of Knowledge, PRINCE2, Agile, etc.), virtual project managers must be proficient in applying the methodologies in a geographically distributed project environment. As an example, communication management is a core competency for any project manager. However, none of the methodologies describes how to apply communication management for virtual (and sometimes multinational) project teams. This knowledge has to be gained outside of core project management methodology training, as does knowledge in all of the core project management competency areas.

Stakeholder Management Skills

Stakeholder management is a skill that is critical to virtual project management success. Virtual project managers may have to manage stakeholders, both internal and external to the organization. Effective stakeholder management helps team leaders gain cooperation from the highly influential stakeholders, cut through competing stakeholder agendas, and positively influence stakeholders who may be inhibiting progress. Project managers who are skilled in stakeholder management must understand three things:

1. Who the stakeholders are and what their needs are.

2. How much influence each stakeholder has on the project.

3. Stakeholder allegiance and attitude toward the project (Project managers should never assume all stakeholders want the effort to succeed).

From this information, virtual project managers can determine which stakeholders must be managed and how to manage them. (See Chapter 4.)

Competency Takes Time

As stated earlier, it is rare for virtual project managers to enter the role proficient in all competency areas. It is also difficult to imagine that any of us could be completely proficient in all areas of the competency model. Review of the model shows why the project manager role requires a broad and varied set of skills. Virtual project managers cannot gain the necessary expertise in a classroom or from a book; competence increases with experience. The role needs to be practiced. Improvement comes with a history of successes and failures associated with actually leading virtual teams and managing virtual projects. Evidence of competence comes from a track record of proven accomplishments that represents an individual's experience base. A considerable amount of time is required to achieve expert-level mastery in anything—including virtual project management.[31]

The knowledge, skills, and abilities described as part of the virtual project manager competency model are, therefore, most useful for growing and developing a firm's virtual project managers, once they are in the role. This capability is important because further gains toward consistent virtual project management success are made through continual improvement in performance.

In ongoing dialogues, direct managers should focus in part on understanding their project managers' growth and career aspirations and should balance those aspirations with management's short- and long-term performance expectations of the project managers. The results from this exercise serve as the basis for an individual development plan for project managers over a period of time (annually in most cases, or more frequently). Career development planning is a process of targeting where individuals currently are in their performance and capability, where they want to be in their career at some time in the future, and then developing a plan on how to get there.

Many project managers are in the discipline because they enjoy the role and want to make it their career. Most highly effective project managers are self-motivated and demonstrate a desire to self-assess, aim for continual improvement, and are persistent in attainment of personal goals.

Virtual Project Management Journey

The situation seems a bit surreal to Jeremy Bouchard as he raises his glass to toast the Sitka project team. He is seated at the VIP table, alongside Sensor Dynamics' senior leaders, who are celebrating the launch of their newest product offering for autonomously driven automobiles and the team that made that happen.

Jude Ames, Vice President of New Product Development, just completed his keynote presentation, a presentation that was full of accolades for the Sitka project team and their performance. The greatest surprise of the evening so far was Ames's words about Bouchard himself: "I would like to extend my personal thank-you to the Sitka project manager, Jeremy Bouchard. Bouchard stepped up to the challenge of leading a project team that spanned multiple sites and multiple continents. He quickly and consistently helped the team perform as a single, cohesive unit." Ames went on to say, "Because

of Jeremy and the entire Sitka project team's efforts, we have nearly $500 million in advance orders for our product. This could be our first billion-dollar product line. I am pleased to announce tonight that the Sitka team has won Sensor Dynamic's highest recognition, the Pinnacle Achievement Award."

Nearly three years ago, as Bouchard was sitting alone in his office, frustrated with his new virtual environment, he could never have imagined his journey as a virtual project manager would lead to this moment.

Becoming an effective virtual project manager *is* a journey. Sometimes, as when a person has no prior experience managing traditional projects, the journey is not overly obvious. However, most of the time the journey is quite obvious and a bit overwhelming because of the complexities of virtual projects. This complexity causes the virtual project manager's work to be more complicated. During the journey, project managers always should rely on and leverage their core project management skills and savviness and remember that leadership is all about people, regardless of where they live and work. Managers should show care, commitment, and concern, along with respect and individualized consideration for every virtual teammate. Doing this provides a good foundation on which to start. Then project managers should layer on the tools, templates, assessments, and best practices detailed in this book to become efficient and effective in managing projects and leading teams in the virtual world.

Notes

1. American Productivity and Quality Center, "Improving Collaboration for Product and Service Development," Final Report, 2008. http://www.apqc.org.

2. Deborah L. Duarte and Nancy Tennant Snyder, *Mastering Virtual Teams: Strategies, Tools, and Techniques that Succeed, 1999* (San Francisco, CA: Jossey-Bass).

3. John A. Pearce and Richard B. Robinson, *Strategic Management: Formulation,*

Implementation, and Control (New York, NY: McGraw-Hill, 2010).

4. Duarte and Snyder, *Mastering Virtual Teams*.

5. Russ J. Martinelli, James M. Waddell, James. M., and Tim Rahschulte, *Program Management for Improved Business Results*, 2nd ed. (Hoboken, NJ: John Wiley & Sons, 2014).

6. Martinelli et al., *Program Management for Improved Business Results*.

7. Stephanie Quappe and Cantatore Giovanna, "What Is Cultural Awareness, Anyway? How do I Build It?" Page 2, 2005. http://www.culturosity.com.

8. John R. Katzenbach, and Douglas K. Smith, *The Wisdom of Teams* (New York, NY: McKinsey & Co., 1999).

9. Alan Brown, *Higher Skills Development at Work: A Commentary by the Teaching and Learning Programme*, Page 4 (London, UK: ESRC, TLRP, 2009). http://www.jfn.ac.lk/OBESCL/MOHE/SCL-articles/Books-chapters-reports/5.Higher-skills-development.pdf.

10. Martinelli et al., *Program Management for Improved Business Results*.

11. Laura Smith, "Global Diversity + Inclusion = Business Success," *Edelman Engage*, Edelman Public Relations. November 21, 2013. http://www.edelman.com/post/global-diversity-inclusion-business-success/.

12. Anne H. Reilly and Tony J. Karounos, "Exploring the Link Between Emotional Intelligence and Cross-Cultural Leadership Effectiveness," *Journal of International Business and Cultural Studies* 1 (February 2009). https://www.researchgate.net/publication/268059609_Exploring_the_Link_between_Emotional_Intelligence_and_Cross-Cultural_Leadership_Effectiveness.

13. Randell Rothenberg, "Noel M. Tichy: The Thought Leader,"*strategy + business* 30 (February 14, 2003). http://www.strategy-business.com/article/8458?gko=87e3c.

14. Dennis J. Cohen and Robert J. Graham, *The Project Manager's MBA* (San Francisco, CA: Jossey-Bass, 2001).

15. James M. Waddell, Tim Rahschulte, and Russ J. Martinelli, "Leading Global Project Teams: Barriers and Challenges", 6 Part Series, *PM World Today*, May 2011.

16. John C. Maxwell, *The 21 Irrefutable Laws of Leadership* (Nashville, TN: Thomas Nelson, 1998).

17. Daniel Pink, *A Whole New Mind* (New York, NY: Berkeley, 2006).

18. Peter Senge, *The Fifth Discipline* (New York, NY: Doubleday, 1990).

19. Jim Keogh, Avraham Shtub, Jonathon F. Bard, and Shlomo Globerson, *Project Planning and Implementation* (Needham Heights, MA: Pearson Custom Publishing, 2000).

20. Martinelli et al., *Program Management for Improved Business Results*.

21. Martinelli et al., *Program Management for Improved Business Results*.

22. Nancy J. Adler, *From Boston to Beijing: Managing with a World View* (Cincinnati, OH: South-Western, Thomas Learning, 2002).

23. Adler, *From Boston to Beijing*.

24. Duarte and Snyder, *Mastering Virtual Teams*.

25. Duarte and Snyder, *Mastering Virtual Teams*.

26. Adler, *From Boston to Beijing*.

27. Cary Chernis, "The Business Case for Emotional Intelligence," 1999. http://www.eiconsortium.org/reports/business_case_for_ei.html.

28. Daniel Goleman, *Emotional Intelligence: Why It Can Matter More than IQ* (New York, NY: Bantam Dell, 1995).

29. Mathew R. Kutz, "Contextual Intelligence: An Emerging Competency for the Global Leader," *Regent Global Business Review*, no. 2 (August 2008): 5–8. https://www.regent.edu/acad/global/publications/rgbr/vol2iss2/RGBR_Vol_2_Issue_2_PDF.pdf.

30. Kutz, "Contextual Intelligence".

31. Malcolm Gladwell, *Outlier,* (New York, NY: Hachette, 2008).

VIRTUAL PROJECT READINESS ASSESSMENT

Throughout this book we have discussed the numerous and complex challenges that virtual project teams face. As a virtual project team leader who is assuming a new virtual project assignment, you will need time and opportunities to periodically assess your team's readiness and ability to succeed. Many managers will view this ability to assess virtual project readiness as a part of their risk mitigation strategy.

The depth and breadth of the readiness assessment needs to be based on the size, complexity, and type of virtual project. Therefore, virtual project managers should utilize those portions of this broad-based readiness assessment that fit their unique organizational and project needs.

Virtual Project Readiness Assessment

Project Readiness	No	In Process	Yes
The virtual project mission and vision are documented and have been approved by the senior sponsors.			
The project goals and objectives are defined, documented, and have been approved by the senior sponsors.			
The project statements of work and scope are clear and documented.			
The project success factors and key performance indicators are defined, documented, and approved by the senior sponsors.			
The project requirements are clear and documented.			
The project requirements are under change management.			
The project business case has been approved by the senior sponsors.			
The project charter has been documented.			
The virtual project team charter has been documented.			
A project complexity assessment has been performed, and associated risks have been identified and documented.			

Project Readiness	No	In Process	Yes
The project assumptions have been documented.			
All project assumptions have been reviewed, validated, or modified.			
Risk events associated with project assumptions have been documented.			
A project budget has been developed and approved by the senior sponsors.			
Adequate funding is available for the project.			
A project schedule has been developed and approved.			
The project schedule had few imposed dates and did not serve as constraints.			
The project schedule has been reviewed and determined to be achievable.			
The project deliverables and key milestones have been mapped.			
Each distributed team or member has a baseline schedule.			
Each distributed project team or member has completed its task plan.			
Each distributed team or member has an estimated budget.			
Each distributed team has a resource plan commitment from functional managers.			
The project risk analysis has been conducted as part of the project plan.			
Project risk identification and management is an ongoing process on the project.			
The integrated project plan is complete.			
The integrated project plan supports attainment of the project goals.			
The integrated project plan has been approved and organizational commitment has been received by the senior sponsors.			
A project communication plan has been developed.			
A comprehensive stakeholder analysis has been completed.			
There are clear decision making processes in place to support the project.			

Project Readiness	No	In Process	Yes
There are processes in place for identifying issues and escalating them to senior managers if they are outside the control of the project team.			
A governance system is in place that accounts for reviewing distributed work progress.			
Common status reporting format and content has been established for the distributed project team.			
The project manager and team use dashboards to track and communicate project progress.			
Project reviews are or will be scheduled on a regular basis.			
All project team meetings are recorded in note form (and immediately posted for review) and in audio recorded form if required.			
No more than two or three weeks separate project team deliverables and outcomes.			

Project Team Readiness	No	In Process	Yes
A project team structure has been established that promotes cross-team collaboration.			
All distributed team sites have been identified.			
All external partners have been identified and are participating in the project planning as required.			
The project mission, vision, and goals have been communicated to the project team.			
The team has a common sense of purpose.			
Team norms have been documented and communicated.			
All tasks have been distributed to the various team locations.			
The experience level of the team is consistent with the complexity level of the project.			
There is low cultural variation within the team.			
Project team members encourage diverse perspectives as a way of performing their project work.			
Cost estimates for team face-to-face meetings have been included in the project budget.			
An initial face-to-face meeting has been scheduled.			
There is team-building training specific to virtual environments and work.			

Project Team Readiness	No	In Process	Yes
There is training on time management and self-management for virtual team workers.			
There is training on asynchronous communication and cross-cultural communication.			
There is a sense of community among virtual project team members.			
Team member profiles are available on the team site or another easily accessible platform.			
A stakeholder influencing strategy has been created for the project.			
Distributed teams or team members use project indicators to track and communicate work progress.			
The team has a dedicated team site to support communication and collaboration.			
Team member roles and responsibilities have been documented.			
The team has the right composition of skills needed to complete the project work.			
Team members are sufficiently proficient in the primary language of the project.			
The project manager has a minimum of two years of experience managing virtual projects.			
The project manager has proven ability to manage cross-team deliverables.			
The project manager has proven to be accountable and meets his or her commitments.			
The project manager has proven ability to make tough decisions.			
The project manager has proven ability to create team chemistry in a virtual project environment.			
The project manager promotes cross-cultural awareness.			
The project manager has proven ability to create project network connections.			
The project manager has proven ability to manage stakeholders from a distance.			

Project Team Readiness	No	In Process	Yes
Team members with delegated stakeholder management responsibility have been identified and trained.			
Conflicts are successfully managed on the project.			
Distributed project team members feel empowered to make decisions and manage work at the local level.			
Team members provide input on fellow team member performance.			
Team members provide input on the project manager's performance.			
A recognition and reward system is in place that promotes team achievement over individual achievements.			
Technology Readiness	**No**	**In Process**	**Yes**
A technology strategy has been completed for the project.			
Virtual communication and collaboration tools have been selected.			
The virtual communication tools selected support the communication usage needs of the project team.			
The collaboration tools selected support the collaboration usage needs of the project team.			
Local infrastructures at the team locations will support data bandwidth and performance requirements for the selected technologies.			
Virtual communication and collaboration tools have been successfully implemented at all team locations.			
All team members have access to the communication and collaboration technology platforms.			
Web-based project management software (or the like) is accessible to all team members at all times to track, post, record, and share project materials.			
All project team members have been trained in use of the technology selected.			
Technology is in place to assist in language translation if applicable.			
All major sites have video conferencing capabilities that the project team can access.			
Project tools promote social connection and interaction among team members.			

Organizational Readiness	No	In Process	Yes
The organizational structure enables cross-organization collaboration.			
The organization has a history of completing projects successfully.			
There is a high level of trust and support among the stakeholders involved with this project.			
There is a high level of change acceptance (low level of change resistance) in the organization.			
There is a high level of leader credibility (of the sponsor and project manager) in the organization.			
The organizational decision making process is participative and collective.			
The virtual project manager is no more than two organizational levels below the senior sponsor of the project.			
The virtual project team members are no more than two levels below the virtual project manager in the team structure.			
The virtual project manager is empowered to make decisions within his or her decision boundaries.			
Virtual team members are empowered to make decisions at the local level.			
The company embraces cultural diversity.			
A core competency and skills model exists for the firm's virtual project managers.			
Virtual project managers have individual development plans that are actively used to increase skills development.			